S0-DUU-593

Sci-Tech Libraries in Museums and Aquariums

The *Science & Technology Libraries* series, Ellis Mount, Editor:

Vol. 6
1/2 Sci-Tech Libraries in Museums and Aquariums

Vol. 5
1 Serving End-Users in Sci-Tech Libraries
2 Fee-Based Services in Sci-Tech Libraries
3 Role of Maps in Sci-Tech Libraries
4 Data Manipulation in Sci-Tech Libraries

Vol. 4
1 Role of Serials in Sci-Tech Libraries
2 Collection Development in Sci-Tech Libraries
3/4 Management of Sci-Tech Libraries

Vol. 3
1 Online Versus Manual Searching in Sci-Tech Libraries
2 Role of Translations in Sci-Tech Libraries
3 Monographs in Sci-Tech Libraries
4 Planning Facilities for Sci-Tech Libraries

Vol. 2
1 Current Awareness in Sci-Tech Libraries
2 Role of Patents in Sci-Tech Libraries
3 Cataloging and Indexing for Sci-Tech Libraries
4 Document Delivery for Sci-Tech Libraries

Vol. 1
1 Planning for Online Search Services in Sci-Tech Libraries
2 Networking in Sci-Tech Libraries and Information Centers
3 Training of Sci-Tech Librarians and Library Users
4 Role of Technical Reports in Sci-Tech Libraries

Sci-Tech Libraries in Museums and Aquariums

Ellis Mount, Editor

The Haworth Press
New York • London

Sci-Tech Libraries in Museums and Aquariums has also been published as *Science & Technology Libraries*, Volume 6, Numbers 1/2, Fall 1985/Winter 1985/86.

The Haworth Press, Inc., 28 East 22 Street, New York, NY 10010
EUROSPAN/Haworth, 3 Henrietta Street, London WC2E 8LU England

Library of Congress Cataloging in Publication Data
Main entry under title:

Sci-tech libraries in museums and aquariums.

''Has also been published as Science & technology libraries, volume 6, number 1/2, fall 1985/winter 1985-1986''—T.p. verso.
Includes bibliographies.
1. Museum libraries—Addresses, essays, lectures. 2. Aquarium libraries—Addresses, essays, lectures. 3. Scientific libraries—Addresses, essays, lectures. 4. Technical libraries—Addresses, essays, lectures. I. Mount, Ellis.
Z675.M94S38 1985 069.95 85-16436
ISBN 0-86656-484-5

Sci-Tech Libraries in Museums and Aquariums

Science & Technology Libraries
Volume 6, Numbers 1/2

CONTENTS

Introduction

Many librarians have little or no awareness of the strong collections and useful services provided by libraries which serve museums, aquariums, botanical gardens, zoos and similar organizations. Many of these libraries are internationally known for their rich holdings and highly professional activities. These conditions hold true for those institutions emphasizing science and technology as well as other disciplines.

It therefore seems appropriate to devote an issue to those sci-tech libraries which serve museums and aquariums so that readers of this journal can be informed about outstanding examples in this category, as well as a survey of some of the less known libraries having sci-tech materials. In due time another issue will be devoted to libraries serving botanical gardens and zoos plus other museums.

In alphabetical order, the issue begins with the library of the American Museum of Natural History, written by Nina J. Root. Following that is the California Academy of Sciences, prepared by Richard L. Pallowick. Our only European entry, Deutsches Museum Library, is described by Dr. Ernst H. Berninger and Eva Reineke. Next comes the Field Museum of Natural History, authored by Benjamin W. Williams and W. Peyton Fawcett.

An academic museum library is described by Eva S. Jonas and Shari S. Regen in their account of the Museum of Comparative Zoology at Harvard University. Following that is the Museum of Science and Industry, written by Carla D. Hayden. Then come two national museums, the National Air and Space Museum, by Frank A. Pietropaoli, and the National Museum of American History, by Rhoda S. Ratner. Next is the Natural History Museum of Los Angeles County by Katharine E. S. Donahue, followed by our sole aquarium entry, the John G. Shedd Aquarium, written by Jan Powers.

A different type of paper concludes the theme-oriented papers, namely a survey of more than a score of libraries serving some of the less known sci-tech museums, compiled and analyzed by Robert G. Krupp.

The first special paper for this issue is devoted to a study of the types of indexing used by sci-tech journals and indexing services, written by Virgil Diodato and Karen Pearson. The other special paper, written by Maryde F. King, deals with the literature written about marketing activities of special libraries in industry. The special collection paper for this issue concerns literature on Alzheimer's disease, written by R. Stephen Hunter.

The issue concludes with our regular features.

Ellis Mount, Editor

American Museum of Natural History Library

Nina J. Root

ABSTRACT. Discusses the history, collections and services offered by the American Museum of Natural History Library. Also described are work on preservation of materials, grant-funded projects and outlook for the future.

The Library of the American Museum of Natural History is generally agreed to have the strongest natural history collection in the Western Hemisphere. Yet for several years after the Library was founded in 1869 it contained only one volume, hardly an auspicious beginning for a Library that was to grow to the point that its present collection contains nearly 450,000 volumes plus countless archives, nearly one million photographs and many other valuable items.

FACILITIES

On April 6th, 1869, the infamous "Boss" (William) Tweed presented a bill to the New York State legislature to incorporate "a museum and library of natural history . . ." in the City of New York. Legend has it that "Boss" Tweed knew what a library was, but Theodore Roosevelt, Sr. and J.P. Morgan had to explain museums to him. Tweed decided that a museum was as good a vote getter as a library and that together they were a winning team. The founders of the American Museum of Natural History (AMNH) also thought that a museum and a library were a winning team and

---

Nina J. Root is Chairwoman, Department of Library Services, American Museum of Natural History, New York, NY 10024. She received a BA degree from Hunter College and the MLS degree from Pratt Institute.

Based in part on the following article: Root, Nina J. Biography of a museum library. *Curator*. 26(3): 185-198; 1983.

promptly set about developing both. While the Museum was at its temporary home in the Arsenal in Central Park, there was no formal library. On November 16th, a few months after incorporation, Albert Bickmore, the founder of the AMNH, presented the German translation of *Travels in the East Indian Archipelago (Reisen in Ostindichen Archipel in Den Jahren 1865 und 1866*, Jena, 1869) to establish the Library formally. This rather insignificant volume now resides in the Library's Rare Book Room, more for sentimental value than for rarity. Eventually Bickmore presented his entire collection to the Library.

According to Bickmore's unpublished autobiography, "The Library was placed in our work rooms where it could be used at any time by original investigators in conchology. This was the beginning of our Museum Library, which then only contained a single volume. . . ." From this meager beginning the Library grew with the addition of donated volumes, reports, and publications from state surveys and scientific societies. Little is heard of the Library until the Museum took possession of its own building on Manhattan Square on December 22nd, 1877. A room "fitted with special cases and shelves of iron was set aside in the attic to house the Library."

Under the leadership of Anthony Woodward, who was named Librarian in 1882, the library continued to grow. In 1911, with Dr. Ralph Tower in charge of the Library and of Museum publications, a public reading room was opened on an experimental basis. That same year an important step was the merger of the Department of Maps and Charts with the Library. By 1916 the Library had expanded to five rooms.

Tower's successor, Ida Richardson Hood, was able to have reading tables and pertinent collections placed in the exhibition halls in 1928 for Museum visitors to use.

No major changes were made in Library facilities during the terms of librarians in the next few decades until 1961, when the Library was moved to its present location, featuring central air conditioning and humidity control. Another improvement took place in 1973 when an air-conditioned Rare Book and Manuscript Room, large enough to house the thousands of rarities needing such care, was built. Over 12,000 rare books have already been transferred to the safety of this room.

Space and proper storage present problems in the Library as well as throughout the Museum. Stacks are overcrowded and the staff is cramped. In 1979 the Library was given additional space in the nearby old Mineral Hall, and the map collection, Museum scientific publications, older serials, and oversized folios were moved, somewhat alleviating congestion. When the Photographic Collection was added to the Library (in 1978), an additional problem arose: the Library and the Photographic collection were in separate buildings, requiring a 20-minute walk. A bridge connecting the two buildings has been completed; now it is a one-minute walk.

COLLECTIONS AND THEIR ORGANIZATION

In its earliest days the Library was heavily dependent upon donations to the collection. For example the first major acquisition of the collection was the 850-volume conchology collection presented by Catherine Wolfe in 1872 in memory of her father, the Museum's first president, John David Wolfe. Other early gifts came after the Museum and the Library moved into the new building in 1878. A month later the annual report for that year announced the addition of a donated collection devoted to ichthyology and other scientific subjects in addition to a set of Ferdinand Hayden's *Reports of the U.S. Geological and Geographical Survey of the Territories, 1867-1878.*

The Board of Trustees and Albert Bickmore, who became the Museum's "Superintendent" at the age of 30, were active in soliciting donations. They successfully acquired such large and important collections as the Hugh Jewett library of voyages and travel, including many rare editions; the ornithological library of Daniel Giraud Elliot, along with the original watercolors by Joseph Wolfe (1820-1899) commissioned for these monographs; the Harry Edwards entomological collection; and the Jules Marcou geology collections. These are but a sample of the extensive donations and purchases that serve as the basis for today's great research collection. The oldest book in the collection is Albertus Magnus's *De Animalibus*, an incunabula, published in 1492 in Venice; the smallest book is Panzer's *Faunae Insectorum Germanica* published in 190 miniscule volumes between 1796 and 1844 in Nurnberg (each volume is about 3 × 5 inches). Systematic purchases in the early years were meager; early annual reports list endless pages of donations and exchanges and only two or three lines of purchases. The Museum publications, the *Bulletin* begun in 1881, *Memoirs* in 1893, *Anthropological Papers* in 1907, *American Museum Journal*, predecessor to *Natural History*, 1900, and *Novitates*, 1921, are greatly responsible for the outstanding breadth of today's 17,000-title serial collection. An exchange program with scholarly institutions throughout the world begun in the early 1880s still serves as an important acquisition tool. It is estimated that some $220,000 worth of material are received each year on exchange.

In the last decade the Library has become greatly involved with nonprint materials. In 1973 some one hundred rare motion pictures dealing with Museum expeditions and history were restored and transferred to the Library, with an additional 3,000 reels added in 1983. Then in 1975 the Museum's collection of memorabilia, consisting of art works, portraits, and mementos of Museum history, were turned over to the Library.

In 1978 the Museum's Photographic Collection of over 800,000 uncatalogued prints, negatives, and slides was placed under the Library's management. Again Albert Bickmore is credited with the founding of this collection. Although photographs were taken and collected from the first days of the Museum, it was Albert Bickmore's lantern slide-lecture series, begun in 1880, that created the first organized photographic collection. Many of his hand-colored lantern slides have been preserved. The Photographic Collection is an extraordinary record of Museum expeditions and scientific history, and researchers and publishers use it heavily. The Museum received a grant to inventory its anthropological photographs and to publish a guide to the collection from the National Endowment for the Humanities in 1980/81. Beginning in October 1981, a project to inventory the balance of the collection, produce a guide, and catalogue the collection began; the Museum received this large two-year grant from the U.S. Department of Education, Title II-C program to Strengthen Research Collections. A computer is used to capture data, sort, merge, and produce indices.

A major three-year project was completed in 1981, funded by a $750,000 grant from the U.S. Department of Education, Title II-C program. The Library was able to join the On Line Computer Library Center (OCLC) international network. Over the three years of the grant project, the Library acquired some 7,000 older titles to fill gaps and added them to the OCLC database. The shelf-list of some thirty thousand titles acquired since 1960 was also entered. The entire 17,000-title serial collection was completely recatalogued and the 250,000 serial volumes inventoried. The information was entered into a database and a computer-produced catalogue was printed. The old serials card catalogue is no longer used and is slated for discard. The card catalogue for pre-1960 acquisitions (the majority of the collection) has been published in 25 volumes plus a supplement. Information contained in card catalogues is being compressed into manageable printed formats through automation. Over the last 4 years the Library has received well over one million dollars in grants! Finally in early 1981 the archives of the Museum were turned over to the Library and moved to a renovated area adjacent to the Photographic Collection.

PRESERVATION AND READER SERVICES

Since the Museum Library has the strongest natural history collection in the Western Hemisphere, its preservation and restoration are of great concern. Several years ago the Museum received a five-year grant to restore the collection. Badly deteriorated volumes were preserved through this grant, and the restoration studio funded by the grant is still used to do repairs. Archives that had been neglected for years are now being cleaned and mended, and then transferred to acid-free preservation folders and boxes. The Photographic Collection's nitrate negatives are being identified and converted to safety film; glass negatives, rare prints, and old albums are being mended, preserved and boxed.

Services are of course not forgotten; they are the prime reason for the Library's existence. The Library primarily serves the Museum scientific staff and circulates over 35,000 items a year among them; about 10,000 members of the public use the Library each year; we answer more than 6,000 reference questions; and as a subject resource for the New York State Interlibrary Loan network we lend about 4,000 volumes to other libraries. The Library is connected to the New York State Library by a TWX that is also used to search the major commercial bibliographic databases; it is an expensive service we are able to provide free to our scientific staff because of our contract with the State Library. Twenty-two permanent staff members provide service, with help from a myriad of part-timers, grant personnel, and volunteers.

OUTLOOK FOR THE FUTURE

Today the collection of the Library is universally acclaimed. It encompasses all natural history subjects except botany, plus excellent reference, history of science, museology, and bibliography collections. The Library has reached a crossroads and must now reevaluate its goals. It no longer needs to acquire older materials, beyond the occasional, hard-to-find retrospective volume. The Library is now ready to capitalize on its great collections. In the future it will use more automation for all phases of library work. We will increase our participation in national and international cooperatives and networks in order to provide the scientific staff with a broad range of collections and services and to make our collections even more accessible to the international scholarly community. We are beginning to address the problem of subject access to scientific photographs and publications dealing with systematics. Also, we must expand the publications program of bibliographies, checklists, catalogues, articles, and books. We have already made a start in this direction. In order to produce camera-ready copy, the Library is entering into a computer bibliographic and subject data for *Recent Literature of Mammalogy* which the American Society of Mammalogists publishes. *Recent Publications in Natural History* is published quarterly; each issue contains one or two book reviews written by the scientific or Library staff. In 1984 the *Photographic Cataloging Manual* was published and has received excellent reviews. Library staff members are frequent contributors of articles in professional journals; one staff member has completed a book manuscript dealing with a rare collection of photographs; another is working on an extensive bibliography on the history of our science; and all the librarians are contributing to, and participating in, national and international library and history of science organizations.

STATISTICS

Name of Museum	American Museum of Natural History
Date Founded	1869
Name of Library	American Museum of Natural History Library
Telephone Number	212/873-1300
Name of Library Director	Nina J. Root, Chairwoman, Department of Library Services

Library Collection Size	
Number of Monographs	156,000
Number of Bound Journals	245,500
Other	Photographs 800,000+; Archives 1,200 lin. ft.; Rare Films 3,000 reels; Rare Books & Ms. 13,000 volumes
Main Subjects Collected	Zoology, Anthropology, Earth Sciences, Paleontology, History & Bibliography of Natural History
Staff Size	9 Professionals 16 FTE Non-professionals
Online searching done	Yes
Interlibrary loans made	Yes
Names of Networks affiliated with	OCLC, New York State Interlibrary Loan Network (NYSILL)

The California Academy of Sciences Library

Richard L. Pallowick

ABSTRACT. Reviews the history of the California Academy of Sciences as well as of the Library. Financial problems, earthquakes and legal complexities are seen to have been surmounted successfully. New developments are also discussed.

EARLY HISTORY

Among the transient multitude of exploited and exploiters in Gold Rush San Francisco was a group who, rather than regarding California as a sort of captive province to be plundered and abandoned, considered it to be both a realm of scientific investigation and home. Seven of them met on the evening of April 4, 1853, and agreed to organize an association to pursue this investigation and to be called the California Academy of Natural Sciences. A week later they drafted the following statement of purpose:

> We have on this coast a virgin soil with new characteristics and attributes, which have not been subjected to critical scientific examination. Sufficient, however, meets the eye of the naturalist to assure him that this is a field of richer promise in the department of natural history in all its variety than has previously been discovered. It is due to science, it is due to California, to her sister States, and to the scientific world that early means be adopted for a thorough survey of every portion of the State and the collection of a cabinet of her rare and rich productions.

Richard L. Pallowick is Assistant Librarian, California Academy of Sciences Library, Golden Gate Park, San Francisco, CA 94118. He has a BA degree from Dartmouth College, an MA degree from Indiana University and the MLIS degree from the University of California at Berkeley.

The constitution adopted on May 16 defined the Academy's object as

> . . . the investigation and development of natural science, the collection of a cabinet of specimens and a library to embrace the standard and current works of natural science, together with such choice miscellaneous literature as may be contributed by the friends and patrons of the Institution.

Such remains essentially the Academy's mission today, although the cabinet has become a major museum, the library has had to withdraw its invitation of miscellaneous literature, and the natural sciences have been thoroughly transformed into science.

None of the twenty-four founding members who incorporated the Academy was a full-time scientist. They were professional men (many of them physicians), businessmen, a few gentlemen of independent means (women were formally invited to join in August, 1853, but none did until 1877). They met after work, in their downtown offices, where they soon incurred the eternal problem of how to maintain adequately organized collections of specimens and library materials in rapidly filled space. A working medical office can sustain only a limited number of bear skulls, pickled snakes, Indian baskets, and German periodicals.

Books were scarce and valuable, and the smaller audience for scientific literature made it more so. Reflecting this importance, the Academy made its librarian an elected officer of the corporation. In one of the most responsible positions, the librarian managed the intellectual capital as the treasurer did the material capital. Both were frustratingly scanty, but the members appear to have shown more concern to increase the former. Their most urgent business was to establish a publications program. When independent publication proved too expensive, the Publications Committee arranged for the *Proceedings* to appear in a religious weekly, the *Sun*, at an affordable rate. The first number was issued September 1, 1854, and consisted of minutes of the previous meeting. The minutes include business affairs, scientific topics covered, and donations lists for specimens, books and even, remarkably, periodical numbers. Does any currently available serials system offer to issue press releases as well as claim forms? Regular publication of the *Proceedings* began in 1873, when everything previously published reappeared and regular original publication began.

While an institution like the Academy must always have been difficult to establish, the economic and social climate of San Francisco presented particular difficulties at this time. The superheated, wildly inflated Gold Rush economy collapsed in 1856; the consequent depression was so severe it reduced San Francisco to chaos and much of the population was threatened by ruin. Among them, Dr. Andrew Randall was shot in the street by a creditor shortly after resigning from the presidency of the Academy in order to devote himself, somewhat belatedly, to sorting out his business affairs. His killer was promptly hanged by the Vigilance Committee (hence "vigilante justice"), a citizens' body which had itself assumed the authority granted to the Committee of Public Safety by the government of the Terror during the French Revolution. Under such circumstances, the perseverance of the founders in keeping alive the Academy amounts to devotion.

Crisis passed into chronic but gradually diminishing hardship for both the city and the Academy. A forecast of worse crisis and greater hardship can be seen in the fact that in 1866 an earthquake cracked the office building in which rooms had been rented to store the collections and library. They were, however, simply stored elsewhere during repairs. The librarian at that time may be the most well-known but indirectly-known individual to hold the office, the geologist Josiah D. Whitney. As the Academy had been active in the campaign to initiate the California State Geological Survey, Whitney soon became a member after he had been appointed head and opened his San Francisco headquarters. He was, perhaps, the first specialized professional scientist, in the modern sense, active in the Academy. Elected librarian in 1861, he also served as president from 1867 to 1869. Mt. Whitney was named for him because he repeatedly predicted that the highest peak in the United States (purchase of Alaska was not yet concluded) would be found in the southern Sierra Nevada. After several surveyors' mistakes in calculation, the highest was correctly identified and finally named.

These were years of quiet growth for the Academy. The Civil War had far less impact than had the earlier depression. Indeed, high demands on mineral resources may have stabilized the local economy, which was still largely gold-dependent, and made easier the Academy's pursuance of its goals. It grew steadily if slowly throughout the period, and established loose, by today's standards, exchanges with similar institutions. The Smithsonian was particularly generous in providing not only its own publications but also duplicate or superfluous material. Exchange agreements, foreign and domestic, became an important resource as the publications program matured; they continue to be vital to the acquisitions budget.

Publications, established to secure appropriate recognition for research, remained a problem. Members repeated complaints that Eastern and European colleagues continued to misappropriate or ignore their work, and sometimes published as their own papers sent them to review. Most vexing, where the first describers had intended to honor Washington by naming the giant sequoia after him—thereby implying he similarly towered above all other historical figures—the British had treacherously alienated the honor to Wellington. The controversy simmered beyond the turn of the century, with neither hero ultimately remaining in the nomenclature.

LICK BEQUEST

Some physical improvements in the storage and organization of all collections did not end these problems. In his annual report of January 1, 1870, the treasurer noted that the Academy was, for the first time, free of debt. But real security would be obtained only through the major philanthropic gift begun in 1873 by James Lick, which was not however realized until 1879. The gap is due, consecutively, to unreasonable restrictions by the donor, to challenges to the will from Lick's illegitimate but recognized son, and to counter-challenges from the Academy. The Academy had been named residual legatee, along with the California Society of Pioneers, and was unwilling to bear the son's tenfold demand upon the estate after Lick's death, with the inevitable diminution of its share. The litigation threatened to bankrupt the Academy but was pursued to the state Supreme Court, which ruled against the Academy. Self-imposed restrictions on its officers' capacity to incur debt narrowed the margin. One of the original specifications that Lick renounced during his lifetime regulated the size and proportions of the library as well as prescribing the overall architectural style (''classic'', a rather backward choice when ''gothic'' already prevailed elsewhere as the standard of seriousness).

Despite the difficulty of procuring it, the Lick Estate contributed enormously to the scientific resources of the Bay Area. In addition to the major donations to the Academy and the Pioneers, it created the Lick Observatory of the University of California. Apparently the Academy's acceptance by local patrons had already risen to new and more generous levels. The library was given rare and important volumes documenting scientific exploration of the Pacific; the institution was enabled to move into the former First Congregational Church (at California Street and Dupont, now Grant Avenue). The first records of paid staff are found at this time. Welcome as the enlargement of space was, the site soon proved inadequate. Dug deep into a hillside of cold wet San Francisco clay, the storage areas were plagued with mildew in zoological and botanical specimens. Part of the collections were opened to the public, free of charge, in 1876.

The most important part of Lick's bequest, was therefore the large, well-located downtown lot on which the Academy began to build two buildings (in Richardson's romanesque style) as soon as the case was settled. One of the buildings fronted Market Street, San Francisco's main thoroughfare, and was designed to be rented out entirely to commercial tenants. The rear building was designed to be the permanent home of the museum, collections, and library and built to higher standards of fire-proofing and earthquake-resistance. The new Academy building opened in January of 1891 to introduce one of the most confident periods in its history. Scientific staff was augmented, expeditions were debated, and even rents discounted to scientific societies for space in the commercial building. After the debates, the Academy purchased and refitted a schooner which it renamed the Academy and sent on a collecting and exploring mission to the Galapagos. The schooner left June, 1905, and was scheduled to return in eighteen months.

1906 EARTHQUAKE AND FIRE

Before then came the earthquake and fire of April 18, 1906. Situated as it was in the heart of downtown San Francisco, the Academy was severely damaged by the earthquake and nearly destroyed by the fire. The earthquake occurred before 6:30 a.m.; by 7:00 officers and staff were trying to save what they could. The first to arrive was President Leverett Mills Loomis, who was soon joined by the librarian (operational, not administrative), Mary E. Hyde, and the curators of herpetology, John Van Denburgh, and of botany, Alice Eastwood, who brought a friend. Together they saved records, type specimens, and whatever else they could find of value. I am particularly indebted to Miss Hyde for saving Theodore Hittell's manuscript history of the Academy, which has been invaluable to the preparation of this article. After no more than three hours all were forced to leave by fire and police officers. The commercial building was destroyed completely; the sturdier museum building remained standing but lost its entire contents. Miss Eastwood, feeling obligated by the greater bulk of the botany specimens, took charge of the material and removed it, first to Russian Hill, then to Fort Mason, and finally back to Russian Hill as the direction of the fire and the forebearance of her friends appeared in her estimation. A letter to *Science*, published May 25, 1906, shows her one of the exemplary survivors of the cataclysm: In it she shrugs off her personal losses and remarks, "My own destroyed work I do not lament, for it was a joy to me while I did it, and I can still have the same joy in starting again." She concludes by reassuring her colleagues of her plans to repair this "great loss to science, and to California", by immediately beginning new collections at type locations.

Mr. Loomis also published a letter in that issue. He and Miss Eastwood both reserve their italics for the loss of the library. This was to be replenished by the good will of institutions around the world. The Smithsonian and Philadelphia's Academy of Natural Science quickly sent letters promising to supply complete sets of their own publications and any additional material they could. Foreign academies similarly sent word, among them those of Austria, Britain, France, Russia and Sweden. The return of the schooner Academy in November, 1906, began the restoration of the scientific collections. It carried specimens so important that they eventually supported two landmark studies in the definition of closely related species groups, John Van Denburgh's description of the Galapagos tortoises and David Lack's *Darwin's finches*. Meetings had already recommenced that September in a temporary location; inevitably, the question of a permanent site recurred.

POST-EARTHQUAKE DEVELOPMENTS

After fruitless decades of contending against the Board of Education for an undesirable site vaguely drawn on a largely conjectural map for a municipal survey, the Academy appealed directly to the city for permission to build in Golden Gate Park. The city referred the question to the ballot; voters approved Charter Amendment 17 in 1910. North American Hall, the oldest Academy building, opened in 1915. All subsequent construction in the park has had to be approved by the city Park Commission, if not the voters. Another referendum item permitted acceptance of a bequest intended by Ignatz Steinhart to establish an aquarium. Opened in 1923, Steinhart Aquarium expanded the educational and research scope of the Academy by allowing live as well as mounted displays. The last major addition was the Morrison Planetarium, which opened in 1952. Planned by G. Dallas Hanna, it had to be designed and built entirely in-house when the second World War cut off the supply of the German lenses (chiefly Zeiss) which had been previously used in the United States.

The Academy continued to grow and develop over the intervening years. In 1950 the ''Executive librarian'' became the ''Librarian'' and the position finally changed status from elective to professional. The library's current housing was opened in 1959 and dedicated to John Ward Mailliard, who had actively sought funds for a new library building during the last years of his tenure as Chairman of the Board of Trustees. Collections were strengthened by a number of significant gifts. The heirs of former President George Davidson presented his collection of sixteenth through nineteenth century exploration literature, assembled while writing *The Pacific coast pilot* guides for western North America. The Belvedere Scientific Fund established by Kenneth K. Bechtel supported acquisition of a comprehensive collection on Baja California, a principal research site for Academy staff. In 1962 a rare book room was built for the Florence and Edward E. Hills Collection, as well as to protect and collocate numerous valuable volumes. The Hills Collection, hundreds of volumes chiefly botanical and ornithological, includes a most appropriate copy of Audubon's Double Elephant *Birds of America*: one saved from the 1906 fire, then owned by the California Institute of Art housed in the former Mark Hopkins mansion. These heavy and unwieldy volumes bear the ''received April 20'' stamps of the San Francisco Police Department. As most accounts of the fire declare the building and its contents a total loss, there is a satisfying symmetry in that an institution whose own library was destroyed should acquire something so nearly lost at the same time.

The archives first received serious attention in 1962, when a volunteer, Mrs. Margaret Campbell, began to seek out and organize relevant material. Increased interest in them lead to the recent appointment of a staff member to administer them.

MODERNIZATION

Automation is also a recent development. An OCLC installation was completed in September, 1983; an M300 Workstation was added in December, 1984 to permit implementation of online searching, serials, and circulation systems. The opportunity to offer an online catalog accessible through the many different personal computers in use throughout the Academy may soon be available, if present software developments continue.

Modernization of library services is only part of current planning at the Academy. A major effort is underway to renovate or replace aging buildings while redesigning exhibits to produce a better integrated, more forceful group of displays. The California Academy of Sciences recognizes the challenge to maintain its commitment to encompass increasingly divergent goals: to pursue advanced scientific research and to provide useful scientific education for the general public.

STATISTICS

Name of Museum	California Academy of Sciences
Date Founded	1853
Name of Library	John Ward Mailliard Jr. Library
Telephone Number	415/221-4214
Name of Library Director	James E. Jackson, Acting Librarian
Library Collection Size	
Number of Monographs	40,000
Number of Bound Journals	70,000
Number of Microform Journal Volumes	100(reels); 100(fiche)
Other	Pictures, more than 1,000,000 images including 100,000 35 mm slides; Maps of Western states; Museum archives

Main Subjects Collected	Systematics & Taxonomy, Natural History of Western states, Botany, Entomology
Staff Size	3 Professionals 26 Non-professionals
Online searching done	Yes
Interlibrary loans made	Yes
Names of networks affiliated with	OCLC

Deutsches Museum Library

Ernst H. Berninger
Eva Reineke

INTRODUCTION

In May 1985 the Deutsches Museum in Munich, Germany, celebrated the 82nd anniversary of its founding. This museum has become known all over the world as an exemplary institution of its kind.

The current arrangement of the buildings on Museum Island, in the riverbed of the Isar, shows clearly the three-part conception for the Deutsches Museum: the largest building contains the collections of historical and contemporary artifacts; in the middle is situated the spaciously planned library building, while the third part is an auditorium. This plan was the concept of the founder of the Deutsches Museum, Oskar von Miller (1855-1934), who felt that the collection of technical and natural science objects should be supplemented by a comprehensive technical and natural science book collection and by the holdings of lectures and congresses. This three-part system is called by its full name: "Deutsches Museum of Masterworks of the Natural Sciences and Technology."

The Deutsches Museum since its conception has continued the tradition of that "Musaion" of which it bears its name and which had been founded in Alexandria in the 3rd century B.C. as a place of cultural interest. That museum also included as its most important part a comprehensive library, primarily built up by Ptolemy II, to serve as the nucleus of a diversified cultural institution. It contained approximately 700,000 scrolls; the same number of volumes is continued in the Library of the Deutsches Museum.

Dr. Ernst H. Berninger is the Director and Eva Reineke is a member of the staff of Bibliothek des Deutschen Museums, Postfach 260102, 8000 Munchen 26, Germany.

Dr. Berninger has the Dr. Phil (Physics) degree from Vienna University and the full academic library degree from "Bibliothekarlehrinstitut, Cologne." Ms. Reineke has been Dr. Berninger's assistant since 1970.

This article was translated by Marlies I. Salamon, a member of the Cataloging Department at Columbia University Libraries.

LIBRARY HISTORY

The founding of the Library of the Deutsches Museum at the turn of the century definitely answered a need. In order to understand this one should visualize the library situation in Germany at that time. The university libraries and libraries of advanced studies were open exclusively to the faculty and the students. The provincial and national libraries, and also to a certain extent the city libraries, had developed from formal royal libraries or from libraries of councils respectively. In them the needs of sovereigns had first priority. The ordinary citizen thus had to overcome a psychological barrier. The church libraries, as far as they could be considered thematically, led a hidden existence behind the walls of convents.

By the end of the 19th century public libraries had emerged here and there, originated by the book-hall movement of Hamburg. But they still were oriented completely towards entertaining, at best offering the belles lettres type of literature. There were no libraries for technically interested citizens; for instance engineers, tradesmen, craftsmen, and others, especially not for juveniles interested in these special fields. Here the Library of the Deutsches Museum was to find its role.

With his characteristic energy, Oskar von Miller turned his plan into reality. In the first year, 1903-1904, the Library received as gifts a considerable number of books, periodicals, autographs, and technical drawings. At the committee meeting on June 28, 1904, Mr. Wilhelm von Siemens reported the following: "For the library collection of manuscripts and drawings to be connected with the Museum, well-preserved bookcases housing approximately 20,000 books have been provided on loan from the library of the Royal Bavarian Army Museum. From approximately 60 donors, from associations, authors, publishers, and private citizens, about 1,500 volumes have been transferred, where the works donated by authors are mostly autographed with handwritten dedications. For the manuscript collection approximately 200 letters, documents, etc., were donated, among them numerous letters from Bunsen, Liebig, Berzelius, Ampere, Humboldt, etc."

At this meeting, the tasks and objectives of the Library of the Deutsches Museum were again outlined by the head of the Institute of Technology at Munich, Professor Walther von Dyck: "We wish to create a scientific central library for technology, mathematics and natural sciences. Along with this a collection of the most important original works about natural sciences and technology from the antiquities must be created and systematically completed for the purpose of historical studies.'' Out of the last mentioned charge grew the department of rare books of the Library of the Deutsches Museum. Professor Friedrich Klemm (1903-1983), who was in charge of the Library from 1921 to 1969, is to be credited with having created a collection of rare completeness. Through special circumstances, the Library did not suffer any losses through the War, so that the collection of source materials today is represented by approximately 5,000 costly volumes. In 1978 a catalog of this department was published.[1]

COLLECTIONS

In the years after the Library's founding, the stock of books grew rather rapidly. In 1908 the Library—which together with the Museum was housed temporarily in the rooms of the Old Bavarian National Museum—already owned 22,000 volumes. These assets had grown fifteen years later, in 1923, to 90,000 volumes.

In 1932 the Library was opened in its new building. Hjalmar Schacht (1877-1970), the then president of the Imperial Bank, had made possible the financing for the library building, in spite of the German inflationary period. The Library building, designed by the architect German Bestelmeyer of Munich (1874-1942), with its impressive entrance hall and the spacious and bright reading rooms for 700 persons, was opened to the public on May 7, three and a half years after its building was started.

The library building was at that time the largest structural steel building in Germany. For the design of the Library, Oskar von Miller went on an exploratory trip to the United States of America and learned there some significant techniques. Especially useful was the placing of the Library's periodicals on the third floor of the building. This turned out to be a special stroke of luck near the end of the Second World War.

The book stack area is constructed as a two-floor stack installation. The ceiling between the second and third floors therefore had to have an especially high load-bearing capacity. From 1943 on, when the bombing attacks increased, the library collections had been transferred to the basement of the library building. The steel/concrete ceiling had the effect of a protecting shield during the heavy bombings. Thus the Library of the Deutsches Museum was one of the few great libraries in Germany which didn't suffer any war losses. By December 1945 the Library could be used in temporarily arranged reading rooms and became a scientific haven for the post-war generation of students in Munich.

At its 50th anniversary in 1953, the assets of the Library had grown to 330,000 volumes. In the following thirty years the Library has continuously developed. Although it lost its preeminence for university students because of the rebuilding of the university libraries and the extension of the library systems at institutes of advanced study, it developed to an increasing extent into being an institution of international research for the history of technology and the natural sciences. Logically, in 1963 an institute of research for these specialties was founded at the Deutsches Museum, in close connection with the Library. In the recommendations on the selection of the scientific equipment, a special chapter was dedicated to the Library. It was already described in statutes as a research library. Since 1945, under the management of Professor Dr. Friedrich Klemm and of Dr. Ernst H. Berninger, it has developed as a free research facility. Thus its acquisitions put special emphasis on source works (originals/second hand, facsimile-prints and editions). The twenty-year-old personal cooperation between the director of the Library and the head of the research institute, from 1963 (founding of the research institute) to 1983, corresponded to the idea of a research and archival library for technological culture.

The transformation of the Library from a heavily visited "reading room" to an intensively used research library has brought along with it some structural changes. The number of reading room seats was reduced in favor of a larger open-shelf section. The Library must in the future take into account the specific necessities for use as a research library. This included the building up of extended services: on one hand there is the need for providing a photocopying department using the latest technical developments, the establishment of a microfilm office, and a provision for using data bases; on the other hand there is a need for serving researchers by an especially qualified cooperative staff. The public rooms must be laid out in such a way that they cater to the needs of the researchers working in them.

At the same time the Library has never neglected its educational charge in connection with the technological and natural science collections of the Museum. Today one could not imagine the educational program of the whole museum without the lectures, seminars and guided tours it offers. They occupy, together with the much-appreciated exhibitions at home and abroad, a central position in the activities of this library.

THE RARE BOOK COLLECTION

Besides its significance as source literature for historical research, the Rare Book collection has an additional function. With its outstanding holdings, it forms a valuable supplement to the perennial exhibitions of the Museum. The Rare Book Department, in connection with the "Writing and Printing Technology" Department, definitely ranks as one of the leading book collections. Its exceptional quality depends upon the fact that in the latter department, the writing and printing technology is demonstrated by pieces of equipment and machines, and that in the rare books collections the products, the manuscripts and books, are presented.

In collaboration with the Research Institute of the Museum a special emphasis on scientific work has been made: the recording, the deciphering, and the conservation of the hot-metal typefaces in the form of typeface samples, trial prints, catalogs and designs.

In special guided tours and lectures, groups of visitors become interested in the development of the natural sciences and technology by familiarizing themselves with original works and facsimile editions.

SERVICE STATISTICS

As the Library of the Deutsches Museum participates in German and international interlibrary loan systems, and thus lends books to other libraries, it can happen extremely rarely that a few books are temporarily not at hand. Otherwise it is a reference library, with no lending outside of the premises.

Hours of business are 9:00 a.m. to 5:00 p.m. daily, including Saturdays and Sundays. Hours of business of the separate special collections and of the educational film service are Monday to Friday, 9:00 a.m. to 5:00 p.m. Telephone: 089-2179-231. The Library is closed on certain days of the year. To be quite sure that the Library is open, it is best to telephone before a visit. The Library is free to everyone.

The Library arranges regular guided tours on each second Sunday of the month at 11:00 a.m. (entrance free). Special guided tours must be arranged for in advance.

Director: Dr. Ernst H. Berninger
Telephone: 089-2179-214
Head of the separate collections and assistant director:
Dip.-Ing. Dr. Rudolf Heinrich
Telephone: 089-2179-220

REFERENCE NOTES

1. Neidhardt, Elske; Nida-Rumelin, M. *Library of the Deutsches Museum. Catalog of printed works to 1750. Rare Books.* Munich: Omnia Mikrofilmtechnik Friedr. Ziffer; 1977. (Series: Catalogorum Bibliothecarum Bavariae Scientificarum, vol. iv).

2. Kataloge der Bibliothek des Deutschen Museums München.
Alphabetischer Katalog. Erscheinungszeitraum: 15.Jahrhundert—1976—Erwerbungsstand: 31.3.1981. 310 Microfiches.
ISBN 3-598-30400-r
Schlagwortkatalog.Erscheinungszeitraum: 15.Jahrhundert—1976—Erwerbungsstand: 31.3.1982. 280 Microfiches.
ISBN 3-598-30402-1.
München u.a.: Saur, 1981.

STATISTICS

Name of Museum	Deutsches Museum, Munchen
Name of Library	Bibliothek des Deutschen Museums
Telephone Number	089/2179-214
Name of Library Director	Dr. Ernst H. Berninger
Library Collection Size	
Number of Monographs	ca. 450,000 vols
Number of Bound Journals	1,200
Other	Special collections
Main Subjects Collected	Natural Sciences and Technology, especially History (18th and 19th century)
Staff Size	15 Professionals 27 Non-professionals

Online searching done	No
Interlibrary loans made	Yes
Names of networks affiliated with	Bayerischer Verbundkatalog (project)

Field Museum of Natural History Library

Benjamin W. Williams
W. Peyton Fawcett

ABSTRACT. Field Museum of Natural History and its Library were founded in 1894. In support of Museum research the Library specializes in the fields of anthropology, archaeology, botany, geology, palaeontology and zoology, all with a global scope. A rich serials collection and numerous special collections make the Library a distinctive resource in the natural sciences. The Library serves both the scientific community and the wider public as a non-circulating reference collection. Through Interlibrary Loan the Library's collections are also available to researchers at virtually all other libraries.

INTRODUCTION

Field Museum of Natural History was founded in 1894 to preserve the collections brought together for the World's Columbian Exposition of 1893. At first the Museum's scope was broad, but in 1896 it began to concentrate solely in the natural sciences of anthropology (including archaeology), botany, geology (including palaeontology), and zoology. Since that time a collection of 9,500,000 specimens in these disciplines has been amassed for scientific study. Each member of the Museum's scientific staff is engaged in one or more research projects either based on these collections or resulting in further additions to them. Field Museum's four scientific departments (Anthropology, Botany, Geology and Zoology) also have research associates who are granted space in the Museum and use of

Benjamin W. Williams is Associate Librarian and Librarian, Special Collections at Field Museum of Natural History Library, Roosevelt Road at Lake Shore Drive, Chicago, IL 60605-2496. He has a BA degree and an MA degree from the University of Chicago. W. Peyton Fawcett is Librarian at Field Museum of Natural History Library. He has a BA degree from Antioch College.

the collections for their research. Work based on the collections is largely published in the four scientific journals published by the Museum: *Fieldiana: Anthropology; Fieldiana: Botany; Fieldiana: Geology; Fieldiana: Zoology.*

COLLECTIONS

The Museum's Library was also founded in 1894 and its collections, now numbering 215,000 volumes, parallel in scope and size those of the scientific departments. Although not specifically designated as such, the Library constitutes essentially a fifth scientific department in the Museum's organization. The Library's holdings are distributed among the General Library (which includes the public Reading Room and Library offices where all processing is centralized); three departmental libraries (Anthropology, Botany and Geology); and six divisional libraries within the Department of Zoology (Birds, Mammals, Insects, Lower Invertebrates, Fishes, and Amphibians and Reptiles). All these libraries are centrally managed by the Library's staff of nine, supplemented by a group of dedicated volunteer workers who assist with both routine tasks as well as special projects. Most of the scientific departments and divisions also have large reprint collections which they maintain separately from the Library collections and with minimal cataloging.

Acquisition of materials from all parts of the world is assured in part through the maintenance of a publications exchange program with approximately 1000 museums, societies, academies and other institutions of great variety worldwide. Over 4000 serial titles are received through the exchange program and more than 1000 through subscription, exclusive of the many hundreds of newsletters, reports and ephemeral periodicals received by various means. Serial publications are the backbone of any scientific library. At Field Museum Library serials constitute 60% to 75% of the departmental and divisional library collections. In the field of entomology, for instance, in which over 400 serial titles are currently received, serials comprise 65% of the Library's holdings. Fifty-six percent of these entomological serials are in languages other than English, typifying the global scope of the Library's collections. Both publications exchange and subscription are essential to the maintenance of such a serials collection. Despite the financial problems that have affected most libraries in the past ten years, Field Museum Library has maintained and even increased serials acquisition in recognition of the importance of these materials for scientific research. Field Museum Library has also been a Federal Depository Library since 1963 and is the only special library (and the only museum library) among the 56 Depository Libraries in Illinois. Several hundred serial titles, as well as other types of material issued by the federal government, constitute a useful addition to the Library collections.

The Library collection is classified according to the Library of Congress system (with some modifications) and uses primarily LC MARC and other records supplied through OCLC. The Library maintains a divided card catalog (Name/Title and Subject) and has revised the Name/Title catalog to conform to AACR 2. The Library maintains several authority files, among them a shelf list, a name authority file, a subject authority file, and a serials cataloging authority file. The Library has been a member of OCLC since 1977 and at the end of 1984 acquired one of OCLC's new M300 Workstations. Not only has the presence of a second OCLC terminal greatly enhanced a variety of processing routines, but the use of applications software in the M300's IBM PC mode has revolutionized the collection and analysis of statistical data, preparation of reports, budgetary control and numerous other tasks.

USER SERVICES

The Library's mission is to support the scientific, educational and exhibition programs of Field Museum and at the same time to make its collections available in a variety of ways to the wider scholarly community and to the public at large. The primary users of the Library are the members of the Museum's scientific staff, research associates and visiting scientists whose research and publication involves extensive use of the Library collections.

Through its Reading Room the Library operates as a noncirculating public library, making its collections available to all who have need of its specialized materials. From 1000 to 1500 patrons visit the Reading Room annually, consulting 8000-10,000 volumes each year. Most of these visitors are students of all academic levels from colleges and universities in the Chicago area who have come to rely on the Library's continuing acquisition of works in natural history, as well as the rich retrospective holdings so important in the disciplines of the natural sciences. Through a variety of formal and informal agreements a number of local institutions and organizations are granted access to the Library's holdings, and have come to rely in particular on the Library's commitment to the acquisition of a thorough serials collection. In addition to these professional and academic users, amateur naturalists and serious hobbyists of all descriptions also make considerable use of the Library collections.

COOPERATIVE PROJECTS

While the Library's holdings have long been known to the scholarly world through the *National Union Catalog, Union List of Serials* and *New Serial Titles*, online access to Library collections became available to members of OCLC when, as noted above, the Library initiated its membership in OCLC in 1977. Although the Library has not entered its retrospective holdings in the OCLC online catalog, all materials received since 1977 appear in the database. Since the Library participates in OCLC's Interlibrary Loan Subsystem, all those materials have become available for loan to member libraries with the ease made possible by online access to holdings information. As a result, and owing in large part to the highly specialized, and not widely held, materials in its collections, interlibrary loans made by the Library through OCLC have increased in number by 40% in each of the last three years. In 1984, for every item it borrowed Field Museum Library loaned 2.3 items to other libraries.

The Library has participated in various cooperative holdings projects in recent years. As a founding member of the Library/Anthropology Resources Group, the Library was involved in compilation of *Serial Publications in Anthropology* (second edition, 1981). Catalog records have also been submitted to the forthcoming *Eighteenth Century Short Title Catalogue.*

SPECIAL COLLECTIONS

Numerous special collections at Field Museum Library contain much rarely-held material. The Ayer Ornithological Library, collected by the Museum's first president, Edward E. Ayer, and widely known through John T. Zimmer's *Catalogue of the Ayer Ornithological Library* (Chicago, 1926), is one of the finest such collections in existence. The Zimmer catalogue set a standard for bibliographic description in the natural sciences and has been reprinted by Arno Press in its *Natural Science in America Series*. Another important special collection is the Berthold Laufer Collection of works relating to the anthropology and culture of China, Japan, India, and Southeast and Central Asia. This large collection was Laufer's working library and includes 1200 works in Oriental languages, most notably in Chinese, Japanese, Tibetan, Manchu and Mongolian. Laufer is widely regarded as one of America's foremost Orientalists and his own works are still considered of the highest importance. Laufer's interests were almost unlimited and his library contains works on far-ranging subjects, from history, art and archaeology to science and technology. His collected shorter papers are being reprinted under the editorship of Dr. Hartmut Walravens. The first two volumes (1976 and 1979) contained over 3000 pages between them, and further volumes are in preparation. The library of George Frederick Kunz was purchased in the year of the Library's founding and formed the nucleus of Library's holdings in the geological sciences. The Kunz collection contains works from all periods in the subjects of mining, mineralogy, gemmology and early works on alchemy, technology and general natural history. The herpetological library of Karl P. Schmidt, a former curator at Field Museum, came to the Library at Dr. Schmidt's death and has been maintained at a comprehensive level by the efforts of the Library and the staff of the Zoology Department. Schmidt's thorough collecting resulted in a virtually complete library on herpetology, consisting of books and an exhaustive collection of reprints.

Certain portions of all these special collections, along with a great many other works, are housed in the Mary W. Runnells Rare Book Room, adjacent to the public Reading Room. The gift of Mr. and Mrs. John Runnells, the Rare Book Room was completed and opened in 1981, and enabled the Library to bring into one modern facility 8000 volumes representing primarily the early literature of the natural sciences. The scientific value of these collections is great, particularly in the fields of botanical and zoological systematics, in which reference to the history of the published descriptions of living things or their fossil remains is essential. For the fields of anthropology and archaeology, the reports of early voyages and travels are often of equally great interest, frequently preserving the earliest descriptions or pictorial representations of European travellers' first encounters with native peoples in all parts of the world. Such works often preserve records of lost cultural traditions while documenting important moments in the early exploration of the world by Western man. One of the most widely appealing attractions of the Rare Book Collections is the great number and impressive beauty of the hand-colored illustrations of the older natural history books. The Ayer Collection in particular is rich in illustrated books and in its entirety provides a virtually complete history of ornithological publication and illustration.

Thanks to continuing support by the Runnells since the completion of the Rare Book Room, the Library has been able for the first time in several decades to make significant additions to the Rare Book collections. The first major addition to the Ayer Ornithology Collection since the death of Edward Ayer in 1927, came with the acquisition of Thomas Brown's *Genera of Birds* (London, 1845), one of the few illustrated bird books Ayer had been unable to locate when building his collection. Also added to the Rare Book collection was a copy of Michael Besler's *Gazophylacium Rerum Naturalium* (Frankfurt, 1642), evidently a proof copy printed on a fine blue paper with startlingly crisp impressions of the copperplate engravings. Most recently acquired, and long a desideratum, was a copy of the first edition of Darwin's *Origin of Species* (London, 1859). No collection such as that housed in the Mary W. Runnells Rare Book Room could be considered complete without a copy of this classic work in the natural sciences.

Field Museum Library has entered a period of reorganization and reevaluation. One of the primary goals of this process is to increase public awareness, particularly in the scientific and library communities, of the resources available in the Library. While there are plans to pursue this goal through a variety of means, at present the Library has in preparation a short title list of the Rare Book Collection and a list of serial titles and holdings. These and other finding tools will greatly enhance Field Museum Library's service to the field of natural science.

STATISTICS

Name of Museum	Field Museum of Natural History
Date Founded	1894

Name of Library	Field Museum of Natural History Library
Telephone Number	312/922-9410
Name of Library Director	W. Peyton Fawcett, Librarian
Library Collection Size	
Number of Monographs	78,000
Number of Bound Journals	137,000
Main Subjects Collected	Anthropology, Botany, Geology, Paleontology, Zoology
Staff Size	5 Professionals 4 Non-professionals
Online searching done	No
Interlibrary loans made	Yes
Names of networks affiliated with	OCLC, ILLINET

Museum of Comparative Zoology Library— The Agassiz Library: Harvard University

Eva S. Jonas
Shari S. Regen

ABSTRACT. The MCZ Library reflects the union between the incomparable nineteenth century natural history values of Louis Agassiz and twentieth century Library and Information Science methodology. Besides maintaining its tradition, which it shares with the Library of the Academy of Natural Sciences in Philadelphia, of being foremost in the field of older historical zoological collections, the MCZ Library has become a respected resource and research center for learning in Zoological and related disciplines world-wide. As part of the Harvard University Library System it supports several teaching and research programs in Biological Sciences offered by the Faculty of Arts and Sciences.

INTRODUCTION

The Museum of Comparative Zoology was founded in 1859 by renowned Swiss-American naturalist Louis Agassiz. Two years later he established the Museum Library through donation of his

Eva S. Jonas is Librarian, Museum of Comparative Zoology Library, Harvard University, Cambridge, MA 02138. She was educated in general zoology and library science at Charles University in Prague; she later received an MA in developmental biology from Harvard University.

Shari S. Regen, also at the Museum of Comparative Zoology Library, Harvard University, is in charge of Special Collections. She holds an MA in teaching English and French as second languages from the School for International Training in Brattleboro, VT, and an MLIS from Simmons College.

The authors wish to express their gratitude to the 1984-85 staff of the MCZ Library for their informative support. Special thanks are extended to Elizabeth Fletcher, Assistant to the Librarian and to the Special Collections for much of the typing of this article on a Macintosh Microcomputer.

own library, help of friends, private subscriptions and purchase of the collection of Laurent Guillaume de Koninck, a noted Belgian paleontologist.

In 1876 the Museum became part of Harvard University. Since then the Library has evolved with the Museum—an outstanding teaching and research Institution of the University.

The Museum has 14 scientific departments, headed by curators who also constitute the teaching faculty of the Department of Organismic and Evolutionary Biology in Zoology, Oceanography and Ecology along with their counterparts from three Botanical Institutions also affiliated with this Department. The Zoology program currently organized by these scientists was rated best in the country in 1982 in an Assessment of Research-Doctorate Programs in the United States, sponsored by the Conference Board of Associated Research Councils of the National Academy of Sciences. Twenty Harvard faculty members are affiliated with the Museum at present and participate in teaching 55 science courses offered by the Faculty of Arts and Sciences. These faculty members along with 5 professors emeriti, 39 graduate students, 10 postdoctoral Fellows, 75 research associates and visiting scholars and 15 curatorial associates and public programs teachers represent the primary users of the Library. The Museum administrators and curators take great interest and pride in their Library. They regard the Library as "one of the Museum's best collections". Many of them support the library by donations of their library collections, books received for review, or journals acquired through memberships.

Several other teaching and research programs depend on the comprehensive MCZ Library collections. Courses offered by the faculty of the Botanical Institutions and the Department of Geology are also supported by the outstanding collections of the MCZ Library, especially its nearly 8,000 serial titles, 1,800 of them current. Undergraduates in biological and pre-medical concentrations use the library's serials collection regularly by their junior year.

As the University's largest science library with 13,500 holdings in 1880, the MCZ library soon established close cooperation with the College Library and later with other science libraries. In 1985 the MCZ Library is still the Harvard's largest science collection of 230,000 volumes, as this parallel process of evolution and cooperation continue. From covering the whole field of natural history in Louis Agassiz's times and most of the first half of the 20th century the collection development policy of the Library became more specialized over the past few decades. At present the library acquires books and serials in Zoological Taxonomy, Paleontology, Ecology, Ethology, Zoogeography, Oceanography and Evolution. The Librarian selects new books and journals after frequent consultations with faculty. Only graduate level research materials with favorable reviews in established journals are regularly purchased. Cooperative acquisitions practices with other Harvard Libraries have existed since the end of the last century thanks to the initiative of Alexander Agassiz, the son of the founder and also the second director of the Museum. Their development reflects new research programs and changes in the structure of the scientific departments affiliated with the Faculty of Arts and Sciences and their libraries.

HISTORY AND NOTABLE COLLECTIONS

Much of the MCZ Library's foundation is based on significant collections that were acquired by gift or bequest over the past 124 years. From its inception, Louis Agassiz enriched the Library's holdings with generous donations from his personal library, including his personal annotated working copies and first editions inscribed by famous authors. Alexander Agassiz (1835-1910), his son and the second director of the MCZ, continued this benevolent tradition by donating his considerable library, including his own publications.

Significant donated monograph, serial and archival acquisitions helped the MCZ Library evolve into a distinguished collection. The earliest acquisition—the library of Laurent Guillaume de Koninck, a Belgian Paleontologist-Chemist, contained various Linnaean first editions.

The extensive entomological collection of the MCZ Library was founded on the library donations of Christoph Zimmermann and Hermann August Hagen. Zimmermann's library, received in 1869, contained many technical works of the 18th and 19th centuries, such as Gyllenhal's *Insecta Svecica* (ca. 1810). His donated library of 762 volumes, acquired (1878-79), included such noteworthy works, as *Fabricius' Entomologica Systematica* (1792).

Outstanding private ornithological libraries were bequeathed to the MCZ Library in the first quarter of the twentieth century. William Brewster, ornithologist and prolific writer, donated his considerable library of 2,093 titles, which included his manuscript journals and photographic scrapbooks in 1919-20. These journals entailed 49 years worth of detailed field observations of New England birds and occupy approximately 30 feet of shelf space in the MCZ Library's Archives. Walter Faxon, who was a Curator of Malacology, donated his personal library of 1708 titles, mostly on Crustacea, which also contained a collection of manuscripts and printed works of Alexander Wilson (1766-1813), "the Scottish father of American ornithology", and a scientific illustrator as revealed in his nine volume work, *American Ornithology* (ca. 1809). Included in the Faxon bequest were such classic works as *Ornithological Biography* (1831) by John James Audubon (1785-1851). In addition, John Eliot Thayer (A.B., Harvard 1885), a loyal donor to the MCZ Library, donated certain manuscripts and well-known paintings by Audubon, such as "The Blackcock" (Lyrurus tetrix) (1827) in 1913-15.

Some of the most estimable books in the Special Collections of the MCZ Library are attributed to the generosity of Samuel Garman, MCZ Curator of Reptiles, Amphibians and Fishes: 1911-24, and bibliophile. Among the 729 volumes and 2,419 pamphlets donated from his library in 1928, is an incunabulum, the oldest treatise in the MCZ Library: *De Animalibus* (1479) by Albertus Magnus ("the Patron Saint of Natural History"). Also included in this extraordinary donation was Konrad Gesner's incomparable *Historia Animalium* (1551-58).

Other donations increased the uniqueness of the MCZ Library's Natural History monograph collection. William McMichael Woodworth, a Keeper of the Museum, donated his collection of 800 volumes that primarily dealt with worms, which was received in 1914-15. Another exceptional acquisition, incorporated into the MCZ Library in 1960, was the collection of Dr. Oliver L. Austen, Jr., on Japanese ornithology. According to the MCZ Annual Report of 1959-60, it included, with the exception of one rarity, "every work of importance on the subject and is rivalled in comprehensiveness by one Japanese institution" (p. 13).

Geological, geographical and travel books are also represented in the MCZ's Special Collections and among its main library holdings. Retrospective acquisitions in this area are exemplified by Hans Sloane's *Voyage to Jamaica* (1707) and Carsten Niebuhr's *Travels Through Arabia* (1790). Josiah D. Whitney, Sturgis Hooper Professor of Geology, donated his geological library to the MCZ in 1896. The personal library of Louis B. Cabot, benefactor, contained books on explorations, given by Henry Bigelow, a former MCZ Curator of Marine Biology, in 1915.

The majority of the rare and scarce donations acquired during the MCZ Library's 124 year development are located in either the Special Collections of the MCZ Library or in the Houghton Library's Storage Area. Book plates revealing the provenance of all donations to the MCZ Library are retained on the fixed endpaper of the inner front board of every volume acquired. In 1941, 377 items of John James Audubon's correspondence were deposited in the Houghton Library of Harvard University. In 1949, one thousand rare books and manuscripts were also transferred to the Houghton Library. In 1979 a huge transfer of geological serials and monographs to the Geology department library was initiated. The monograph transfer is still under way. The Special Collections Section of the MCZ Library was officially instituted in 1978, although a special section of the library classified "X" existed in locked cabinets from the beginning.

Excluding the five museum staff members entrusted with the MCZ Library in its earliest stage (Alexander Agassiz, Jules Marcou, J.B. Perry, F.R. Staehli and P.R. Uhler), there has been a succession of ten Librarians from 1871, to the present day. In consecutive order, they have been: Ms. Frances Slack, Mr. Samuel Henshaw, Mrs. Eleanor S. Peters (a.k.a. Eleanor K. Sweet), Mr. William E. Schevill, Mrs. Margaret A. Frazier, Mr. Robert L. Work, Ms. Jessie B. MacKenzie, Mr. C. Munetic, Ms. Ruth Hill, Ms. Eva S. Jonas.

SPECIAL COLLECTIONS

During the 124 year history of the MCZ Library, valuable books, archival ephemera and artifacts have been acquired. The Special Collections section of the MCZ Library was officially instituted in 1978. In the following years starting with the published G. K. Hall catalog, all records were reviewed, rare books and serials were annotated and then incorporated into a computer based short title catalog. Several thousand volumes were added to the Special Collections area, where the X collection from locked cabinets and volumes from earlier transfers were shelved. In 1981 an electronic security system for this area and the main entrance of the library was installed. The Special Collections with closed stacks and a reading room, houses 15,650 volumes (dated from 1478 to 1983). On rare occasions, a new, limited edition may be deposited directly into the Special Collections, as was the case with Forshaw and Cooper's boxed folio of *Alcedinidae* (Kingfishers) (1983). Part of this collection, about 1,000 volumes, is retained in the Houghton Library, where the MCZ has an allocated deposit. Primary access to the Special Collections is via three printout aids arranged by author, subject and chronological fields from the computer short title list. Other special printouts have been produced from the catalog to increase security of the rare books as well as their records and to facilitate their inventory. A complete inventory of monographs and serials in Special Collections was undertaken in 1981 and again in 1984.

Any patron who intends to enter the Special Collections for use of its monographs, serials or archival materials, must first sign a guest book, must leave his/her possessions, excluding pertinent citations, at the Circulation Desk and must press a buzzer for admittance. Rare materials are being used in a supervised Reading Room.

RECORDS, CATALOGING, CLASSIFICATION

The cataloging system was designed by Jules Marcou in 1862, when he and Alexander Agassiz cataloged the whole collection. Monographs are divided into subject areas, for example, Aves, Mammalia, Pisces. Within each subject area, they are arranged alphabetically by the authors' last name. Multiple publications by one author are listed in order of publication date, in a chronological sequence, rather than alphabetically by title. The accession number consists of the general subject area, followed by the first letter of the author's surname. This cataloging system was used between 1862 and 1977.

J.B. Perry, an Assistant Curator of Paleontology, used the sales catalog of the de Koninck library as the Accession Record for the first 3,015 volumes. Subsequent meticulously hand-written Accession Logs would be continued through 1947. A shelf-list catalog was begun by Mr. Samuel Henshaw, Curator of Entomology and the eventual third Director of the MCZ.

In 1941, the project of typing catalog cards began. In the MCZ Annual Report of 1941-1942, W.E. Schevill reported that on October 1st a crew of trained typists under the direction of Miss Margaret Currier of the Harvard College Library began copying the old manuscript catalog on 3″ × 5″ cards in accordance with the standard Harvard system (p. 40). Copies of MCZ Library cards were filed in the Union Catalog at Widener Library. The MCZ Library Card Catalogue was published by G.K. Hall & Co. in 1967 in eight volumes. The publication has facilitated the coordination of activities in the various natural science libraries in the country and has proven to be a much consulted bibliographic aid.

Since 1978, all monographs received by the MCZ Library have been placed into the Library of Congress classification. All reference works, including those in the Special Collections, have also been reclassified into the L.C. system, i.e., "REF". The Permanent Reserve ("PERM RES") collection and the Ready Reference section also located in the Reference Room have all been recataloged and reclassified. The Library participates along with other Harvard libraries in the OCLC cataloging system. All records since 1978 are part of the Distributable Union Catalog available in all Harvard libraries on microfiche. A special computer-based short title catalog of the rare book collection was produced in 1981.

SERIALS AND THEIR CLASSIFICATION

The collection of serials takes more than 70 percent of the Library's stacks space and 80 percent of the current acquisitions budget. It contains 8,000 serials titles, 1,800 of which are current. Two thirds of those are received through an exchange program of Museum publications initiated by Louis Agassiz. Publications of Museum papers, incorporated into the leather series, begun in April 1863, were sent out in exchange to prominent institutions, "both home and abroad" for further acquisitions, especially in regard to Learned Societies' Transactions, as indicated in the 1867-68 Annual Report. The *Bulletin* of the Museum of Comparative Zoology, *Breviora* and *Psyche* are currently sent out to all parts of the world.

An accessible, alphabetized system for most of its serials was adopted by the MCZ Library in 1980-81. All current serials and complete runs of dead serials are included in this system. These serials were recataloged and incorporated into the CONSER database and simultaneously into the Distributable Union Catalog of the University. These journals are arranged alphabetically by title, unless the name of the issuing institution is part of the title. The call number is composed of the first three letters of the title in capitals followed by a four digit number that corresponds numerically to the alphabetical sequence. For example, *Nature* magazine is accessed as NAT 5064. If the name of the issuing society is integral to the title, the call number begins with the first three letters of the issuing society and the journal is classified according to the societal name. This is exemplified by the serial, *Proceedings of the Zoological Society of London (ZOO 8613.5a).*

Other serials are kept in the old MCZ classification system. Serials in a particular subject area are grouped together with a corresponding call number, consisting of the monographic classification code, followed by a lower case "j" and the first letter of the title, e.g., Aj = Aves (Birds), Pj = (Pisces) Fish, Pzj = Paleonzoology. Entomology journals have the call number: E.D., without a "j", followed by the first letter of the title and Marine Science journals are interfiled with Oceanography monographs. Furthermore, serials issued by institutions are classified by the geographical location of the institution: first, by continent and next, by city. If an institution publishes more than one periodical, the series is alphabetically arranged on the shelves. These are represented by: (S-AF: Serials—Africa, S-Au: Serials—Australia, S-ES: Serials—Europe/Siberia, S-IA: Serials: Indo-Asia, S-NA: Serials—North America and S-SA: Serials—South America). An example is "Steenstrupia," issued by the Zoological Museum of the University of Copenhagen, classed in S-ES-C[openhagen].

Some journals are classified alphabetically. Their accession numbers consist of "S-," followed by the first letter of the first word in the title, e.g., *Asiatic Journal* (S-A 832). This older system is more prevalent in the Special Collections.

There are several finding aids for locating serials in the MCZ Library. In order to find the call number of a journal, patrons are recommended to look in "The Index to MCZ Serials", which is a computer-generated printout. This unique MCZ Serials Index is comprised of titles, all their possible variations and abbreviations, and their issuing bodies, in an alphabetical arrangement, which includes the most often used serials, as well as all current ones. If the journal is not listed in this "Index", it may have been classified according to the older MCZ systems. These might then appear in a Circular File, in which periodicals are listed by title only. Serials, originally filed in the older MCZ arrangement, are also listed in the Card Catalog by title, issuing society, and, in some cases, by geographical location. This Catalog also provides a summary of holdings of the journal. Additional printouts of "Periodical Publications in the Harvard Science Libraries" list periodicals by title and by keyword in the Harvard Science Libraries. In addition, the "Harvard University Library Serials List," which is arranged by title and which includes non-science libraries, can also be consulted in MCZ serial searches. When a periodical that was issued within the past few years is needed, users are advised to check unbound journals in the Agassiz Room, as well as the Bindery File at the Circulation Desk. For further reference, the Reference Librarian can check several other reference sources to locate the journal in other libraries. In order to promote independent access, there is also a slide/tape guide entitled, "Searching for Journal Articles in the MCZ Library".

POLICIES, SERVICES AND PROCEDURES

Upon entering the MCZ Library, everyone is obliged to show identification. Non-Harvard users have to register at the Circulation Desk. A friendly attitude to the public interested in Natural History is another tradition maintained since Louis Agassiz's times. If unaffiliated with University, the first visit to the MCZ Library is gratuitous. Subsequent use requires membership payment, with a special arrangement for students and faculty from other institutions. In 1985, the MCZ Library is open Monday-Friday, including three nights and Saturdays (10-1). Graduate students have proved to be regular evening patrons.

Interlibrary Loan is another service offered by the MCZ Library to officers and graduate students of the Museum. I.L.L. requests may be deposited at the Circulation Desk or with the Interlibrary Loan Assistant. The user incurs all charges and may ask for an estimate and/or indicate the maximum accepted charge on the request form. There are reciprocal agreements between the MCZ Library and several other libraries for loans and photocopies without charge. It has been perceived over the past two decades, that more I.L.L. requests have been received from Japan to the MCZ Library and from the Museum Library to South America.

Special aids assist users in the MCZ Library. Microforms are available and may be accessed via microfiche and microfilm readers in the Reference Room. There is a small microfiche collection in the

Special Collections, as well, which serves to replace missing volumes and to augment holdings. There are slide tapes to facilitate use of the following scientific indexes: *Science Citation Index, Biological Abstracts* and *Zoological Record*. Moreover, there is a taped tour of the MCZ Library, which is a graphic, informative introduction to the Museum Library. Each Fall, since 1978, the Librarian has implemented Bibliographic instruction with the use of these slide tapes for MCZ Graduate Student orientation.

Technology has enlightened the MCZ Library since the early part of the twentieth century. In 1927, electric lights were installed in the stacks and in the reading rooms. In 1984, the MCZ Library took an automated leap when a Macintosh Microcomputer with peripherals, was acquired. Its software is frequently used: MacWrite software is used for all correspondence, lists and word-processing tasks. Its MacPaint software is used for signs, maps, blueprints and graphics. The Habadex software has been adopted as a database file system for creating temporary records of donations and additions into the Special Collections. By 1985 another Macintosh had fortunately been acquired. The MCZ Library staff and the Librarian are finding more use every day for this user-friendly Apple Microcomputer.

During the past year, the MCZ Library has been incorporated into HOLLIS (Harvard On-Line Library Information Service). Consequently a HOLLIS terminal was recently installed in Technical Services. It is primarily used for automated serials acquisition control and for revising and updating the format of serial holdings.

CONCLUSION

From its inception as a one-room, one-Librarian Museum Library of circa 6,000 volumes, the MCZ Library has grown into the estimable collection of approximately 230,000 volumes of monographs, and journals and of archival materials. It now has 18 rooms and 21 staff members: 9 full-timers (3 Librarians and 6 Assistants, 4 Casual Workers and 8 Student Aides). The MCZ Library has witnessed a century of transition: in evolutionary thought, in bibliographic control, in preservation practices and in information technology. In 124 years, 10 successive Librarians have contributed to the updating and classification of the large and important collections donated to and bought by the Library. Many of them have augmented the MCZ Library, vis-à-vis collection development, administrative policy and physical expansion. The MCZ Library reflects the union between the incomparable nineteenth century natural history values of Louis Agassiz and twentieth century library and information science methodology. Besides maintaining its tradition, which it shares with the Library of the Academy of Natural Sciences in Philadelphia, of being foremost in the field of older historical zoological collections, the MCZ Library has become a respected resource and research center for learning in zoological and related disciplines world-wide. As part of the Harvard University Library System it supports several teaching and research programs in biological sciences offered by the Faculty of Arts and Sciences.

REFERENCES

Agassiz, A. *The Harvard University Museum—its origin and history.* Cambridge, MA; 1902 June 12: p. 13, 15.

Assessment of research-doctorate programs in the U.S. biological sciences. Committee on an Assessment of Quality-Related Characteristics of Research-Doctorate Programs in the U.S., National Research Council; 1982: p. 193.

Museum of Comparative Zoology. *Annual Reports.* Cambridge, MA: 1867-68, 1876-77, 1927-28, 1941-42, 1943-44.

Museum of Comparative Zoology. Library Staff. *A Guide to the Use of the M.C.Z. Library.* Cambridge, MA; 1975 September.

Museum of Comparative Zoology. Library Staff. *M.C.Z. Library Guide.* Cambridge, MA; 1983.

O'Hara, R.J. *Principle private book collections now in the library of the M.C.Z.* Cambridge, MA; 1983 July. Unpublished.

Scudder, S.H. *Scientific institutions of Boston and vicinity.* Boston, MA, 1880; p. 15.

Work, R.L. Professor Agassiz's Natural History Library. *Harvard Library Bulletin;* VI (2):202-218; 1952 Spring.

STATISTICS

Name of Museum	Museum of Comparative Zoology, Harvard University
Date Founded	1859
Name of Library	Museum of Comparative Zoology Library
Telephone Number	617/495-2475
Name of Library Director	Eva S. Jonas, Librarian

Library Collection Size	
Number of Monographs & Bound Volumes	230,000
Number of Microform Journal Volumes	235(reels); 4580(fiche)
Other	Museum archives, 300 lin. ft.
Main Subjects Collected	Zoological Taxonomy, Paleontology, Ecology Oceanography, Evolution
Staff Size	3 Professionals 10 Non-professionals
Online searching done	No
Interlibrary loans made	Yes
Names of networks affiliated with	OCLC, CONSER

Museum of Science and Industry Library

Carla D. Hayden

ABSTRACT. Describes the history of the library, its services, its collections and its facilities. Its emphasis on educational and children's materials is discussed.

INTRODUCTION

Daily library service to the public is once again a feature in the Museum of Science and Industry. The Museum's library has been remodeled and reopened after over 40 years. Library service at the Museum reflects many aspects of the growth within the nation's oldest, largest, and most popular contemporary science and technology museum.

The Museum is Chicago's leading tourist attraction, with an annual attendance close to four million. There are 75 exhibit halls dealing with the physical and life sciences, engineering, and mathematics. The Museum has more than 2,000 exhibit units that are three dimensional and participatory, designed to be operated by visitors. The emphasis is not on collecting artifacts but on graphically illustrating scientific advances and their technological applications.

The Museum is housed in the only remaining structure from the World's Columbian Exposition of 1893, in Jackson Park at 57th Street and Lake Shore Drive, on the city's South Side. The initial funding and impetus for the Museum's founding came from philanthropist Julius Rosenwald, who had been greatly impressed by the Deutsches Museum in Munich. After several years of planning and reconstruction, the Museum of Science and Industry was opened to

Carla D. Hayden is Library Services Coordinator at The Museum of Science and Industry, 1700 East 56th Street, Chicago, IL 60637. She has a BA from Roosevelt University and an AM from the University of Chicago.

the public in 1933 during the Century of Progress Exposition. At that time only a few exhibits were prepared and the library was opened but in temporary quarters.

EARLY YEARS

The Museum library was organized as early as 1928 when headquarters for the entire Museum were established. Records of library service from 1928 to 1940 remain. By 1938, the library had approximately 25,000 volumes with extensive trade catalog and pamphlet collections. Also during that year, 18,000 museum visitors and researchers used the resources in the temporary library quarters. The permanent facility opened in 1939, and was in the central section of the museum with a separate children's room. The library was open daily for reference use by any museum visitors.

The full opening of the major portions of the Museum was completed in 1940, along with a major shift in administrative direction. When Lenox Lohr became the Museum's president and director that same year, one of his first acts was to dismiss several curators and the head librarian. The previous director, also dismissed, noted "Mr. Lohr's dictum to close the library, his dismissal of the librarian, is in strange contrast to the enlightened Alexandrian practice", and "the library had particular appeal to Mr. Julius Rosenwald". Yet, the dismissals were carried out and the library closed to the public except by appointment only. The primary focus of the library changed from public and staff services to staff support only. This policy continued throughout the next decade as the Museum's focus changed as well. The paucity of records and correspondence reflect the irregularity of professional library staffing during that period. The next mention of a full-time librarian is in correspondence from Lohr in 1955, stating that, "the library was closed for a while last year because we did not have a librarian".

The newly assigned librarian, (service from 1952 to 1972), completed a 1954 Greater Chicago Library Directory listing one staff member, 20,000 books, 300 periodicals, and reference service by appointment only on weekdays. As the popularity and status of the Museum grew, interest in the library was stimulated for public usage. An important aspect of the library's collection remained the Seymour Dunbar collection of 2,000 prints on transportation.

From 1952 to 1959, inquiries for use of the library were handled through the Manager of Operations. The library's main function, in a 1960 reply for service, was "to supply our museum demonstrators with reading and studying materials explaining the exhibits". It was also noted that "ours is not a very active library". Since 1940, the juvenile portion of the library had been discontinued completely, though a collection of children's materials was retained. By 1971, there had been no real collection growth or change in scope. Purchases were primarily on recommendation of staff members and many volumes were on permanent loan to various Museum departments. The staffing of the library did not exceed two staff persons after the 1940 closure. The library had become an underutilized and stagnant facility.

EXPANSION OF SERVICE

The present Museum director and president, Victor J. Danilov, appointed in 1972, initiated the concept of expanding and renovating the library into a center of science education with open access for all visitors of all ages. An education department was created to provide quality supplementary, informal educational experiences and a new library was envisioned as a vital part in the new department's activities. The library's collection of juvenile literature, though not readily accessible to children, had been enhanced by Museum participation in the Chicago Tribune's annual book fair, "Miracle of Books", from 1953-1968.

In 1972, the Museum began sponsorship of the Children's Science Book Fair, which provided an excellent core collection of science materials for youth.

The initial feasibility study and preliminary design concept (1976) for funding of the proposed multi-resource learning facility put emphasis on a new children's science library, called a Student Science Center, that would be "a place to come to be stimulated, assisted, and possibly find answers".

NEW FACILITY

A proposal was submitted to the Kresge Foundation in 1981, to "develop an innovative children's science library that will be the first of its kind in the world". The proposed library facility was to be built in the same location as the previous one. The Kresge Foundation granted a challenge sum and the new library became a high priority item on the list of funding opportunities for the Museum's 50th Anniversary campaign. By late 1982, the majority of funds were received. The plans for the new facility included the Children's Library, a teacher resource area, and sections for casual Museum visitor browsing. The Kresge Library was formally dedicated on December 8, 1983, and opened to the public on January 2, 1984. The library is part of the Museum's Education Department which includes the Seabury Laboratories, programming activities, teacher and school group services, preschool exhibit and computer laboratories.

The Kresge Library houses a special collection of materials regarding science and its related subjects for children, adults, and educators. The 10,496 square feet of space contains electronic and audio-visual media for study and review, appropriate hardware, and resources in a variety of formats. The library's collections are for reference use within the facility and are available for in-house usage to patrons of public libraries throughout the state via interlibrary loan agreements.

The library's collections consist of books, periodicals, pamphlets, activity kits, films, cassettes, filmstrips, videotapes, slides, computer software, maps, posters, textbooks, learning games and toys, information and curriculum files, and puzzles.

The collections of materials are divided physically and organizationally into:

1. a collection of materials for youth, preschool to senior high school level;
2. materials for teachers and parents to aid in teaching science to those groups;
3. a general popular science collection of books and periodicals for adults and Museum staff members.

The library is open daily for children seeking resources for science projects, participants in Museum classes, parents and teachers who want to work more effectively with their children and students, and any Museum visitor investigating exhibit subjects.

The architectural firm hired for the renovation project combined soft forms, muted colors and indirect lighting to achieve the Art Deco style of the 1930's. The effect is carried throughout the facility, with variations in furniture style and color patterns.

The main entrance to the library is on the Museum's first balcony level. Upon entering, the visitor can seek general information at a low level desk and browse through periodicals and a changing selection of general interest books displayed face-out, in bookstore style. Comfortable, informal seating contributes to the relaxing atmosphere of this browsing area. On this level the library is configured in an "L" shape, with a 2,303 square foot youth section directly connected to the entry browsing area. The youth section of the library is divided structurally to provide areas for preschoolers at one end and older teens at the other. At the preschooler's end, there is a free-standing mini-theater and shelved semi-circular unit filled with picture books (230), puzzles and activity boxes. The chairs and table are scaled to size with an adjacent section for parents. The parents' section includes comfortable seating with extensive child development materials. Preschool story hours on science concept themes take place in the mini-theatre area. School groups can also view filmstrips, films or receive orientation in this area. Individuals may look and listen to audiovisual materials along a counter in the back of the room.

At the other end of the room is a partially enclosed computer learning section where young people can use the library's collection of computer software on five microcomputers. Periodicals, books, and other materials relating to computer technology are available nearby. Along the outer wall areas of the youth section are special focus units containing books on subjects such as the human body, space, energy, and science projects. The focus units are supplemented with posters, globes, simple in-house projects, and study prints. One of the units focuses on careers in science and related occupations with an emphasis on women and minorities.

The core of the children's collection is shelved in the middle section of the room. The entire youth collection is approximately 3,700 volumes.

The second level of the library houses the general adult science collection and teacher resources, such as textbooks, reports, trade catalogs, and education periodicals. When the renovation project was confirmed, 75 percent of the adult collection was sold to a southwestern university. The Dunbar transportation prints were sold as well.

At the Interlake Science Service counters teachers can review commercial and Museum produced activity classroom kits. Microcomputers are available for educators to preview science software for classroom use. Filmstrip viewers, slide viewers, and videocassette equipment are accessible for reviewing audiovisual materials. Teachers will find laminating equipment, a photocopier, and slide and filmstrip production machines for making original materials to take back to their schools.

The library will have a NASA Regional Teacher Resource Room with a collection of slides, filmstrips, and videocassettes relating to NASA space programs. These materials can be reproduced for the cost of materials on the library's reproduction and duplicating equipment. The NASA resources also include teacher curriculum packets and handout sheets. The materials cover all areas of NASA activity, including the manned space program, interplanetary probes, and aeronautics. Most of the materials can be easily incorporated in classrooms but are not generally included in current textbooks.

The third level of the library has back issues of over one hundred periodicals and a rare book room with the Museum's collection of World's Fair items and other historical materials.

The Kresge Library utilizes the Library of Congress System for all monographs; the original portions of the collection were also cataloged in Library of Congress.

The Kresge Library is participating in a cluster of libraries in the Chicago metropolitan area that share a main frame computer to provide touch terminal access to the libraries' collective holdings. The Museum library's participation not only provides access to the bibliographic records of the cluster's 48-member libraries but also dial-up retrievability to the seven other state systems, including the Illinois State Library, utilizing the same automation technology. There are touch terminals in the browsing area, youth section, and on the second level. The library is an affiliate member of the Chicago Library System. Plans are underway to gain access to the Chicago system's database.

There are three professional staff members, two part-time library aides, and one clerk typist. The majority of the staffing resources is made up of dedicated volunteers, numbering 25. Their contributions include data entry, technical services, public service, NASA material duplication, research and development. The ages, interests and backgrounds of the volunteers are indicative of the Museum's diverse audience.

The primary publication of the Kresge Library is *Children's Science Books*, an annotated bibliography of books from the Science Book Fairs.

STATISTICS

Name of Museum	Museum of Science and Industry
Date Founded	1933
Name of Library	Kresge Library
Telephone Number	312/684-1414
Name of Library Director	Carla D. Hayden, Library Services Coordinator
Library Collection Size	
Number of Monographs	16,000
Other	2700 Software programs; 110 Videocassettes; 1000 Slides; 40 Filmstrips, kits, maps, textbooks
Main Subjects Collected	Physical Sciences, Life Sciences, Education, Technology, Museology
Staff Size	3 Professionals 3 Non-professionals
Online searching done	No
Interlibrary loans made	Yes
Names of Networks affiliated with	Chicago Library System, Suburban Library System/ CLSI Cluster

National Air and Space Museum Library

Frank A. Pietropaoli

ABSTRACT. The National Air and Space Museum Library, a branch of the Smithsonian Institution Libraries, provides materials and services to support the varied research programs of the National Air and Space Museum. Brief histories of the Museum and of the branch library and a summary of Museum programs provide a background for an overview of current library users, resources, and services.

INTRODUCTION

The Smithsonian Institution Libraries (SIL) is a dispersed system of branch libraries with centralized administration. Acquisition, technical processing, and administrative services are carried out in centralized units while the public service units are distributed in various Smithsonian Institution museums and bureaus to support the specific research programs of those units. The National Air and Space Museum Branch Library (NASM Library) is charged with developing and providing the resources and services needed to support the varied programs of the National Air and Space Museum (NASM). Consequently, in order to appreciate the role of the NASM Library, it is necessary to be aware of the mission and programs of NASM. This paper gives a brief history of the Museum, summarizes Museum programs, identifies library users and their needs, and describes library resources and services.

THE MUSEUM AND ITS PROGRAMS

Since its establishment, the Smithsonian Institution has had an interest in subjects such as astronomy, natural flight, and ballooning, all of which are current interests of NASM. Towards the end of the

---

Frank A. Pietropaoli is Chief Librarian, National Air and Space Museum Branch Library, Smithsonian Institution Libraries, Washington, DC 20560. He has the BA degree from Union College and the MLS degree from Catholic University.

nineteenth century, accelerated experimentation with heavier-than-air craft stimulated interest in the scientific and technical aspects of mechanical flight. Samuel Pierpont Langley, Secretary of the Smithsonian Institution from 1887 to 1906, was himself actively engaged in such experimentation. Following the success of the Wright brothers in 1903 and the rapid developments thereafter, man's conquest of air captured the imagination of the scientific and technical community as well as the public at large. The association of Robert Goddard with the Smithsonian which began in 1916 similarly stimulated an interest in rocketry. As the twentieth century progressed, the fantastic developments in unmanned and manned spaceflight further enlarged the scope of Smithsonian interest and research.

As a result of the scientific and technological developments summarized above, the Smithsonian Institution acquired aerospace artifacts which grew in number and complexity. Until the establishment of a separate museum for these collections, they remained curatorial collections in existing departments of the United States National Museum.

The establishment of the National Air Museum was authorized by law in 1946 to memorialize the development of aviation; to collect, preserve, and display aeronautical equipment; to serve as a repository for scientific equipment; and to provide educational material for the study of aviation. Public Law 89-509 (1966) changed the National Air Museum to the National Air and Space Museum. This law expanded the Museum's mandate to include collection and research on space flight as well as aviation and, further, authorized the construction of a new building to house exhibits and the research staff. Since the July 4, 1976 official opening of the NASM building on the Mall in Washington, DC, there have been well over 75 million visitors, making it by far the most widely visited museum in the world.

Physically, NASM comprises the current museum building on the Mall plus fourteen large buildings at the Paul E. Garber Preservation, Restoration and Storage Facility located in Suitland, Maryland. In addition to conducting programs indicated by its name, the facility is used for exhibit production and for the display of aircraft and other flight-related objects. The NASM artifact collection numbers over 35,000 pieces including large aircraft and spacecraft. Even with a large museum building and several display buildings at the Garber facility, only a limited number of artifacts can be exhibited at any given time.

NASM LIBRARY BACKGROUND

As aerospace artifacts were acquired by the Smithsonian Institution, the researchers collecting and working with the artifacts required printed materials to document the collections. These materials consisted of technical documents, histories, biographies of important personalities and the like, which not only permitted research relating to the technical aspects of the artifacts for restoration and exhibition but which also placed the development of these artifacts in proper historical perspective and sequence.

As with most Smithsonian Institution Library Branches, the collections of documents and books began as the very modest working collections of Smithsonian curators and other researchers. These private collections grew as the artifact collections grew. As research staff left the Institution, retired, or died, the printed materials were left with the Institution. Eventually, they grew to a size to require separate housing and became a "library". The library collections were further enlarged and expanded through donations of material from private individuals and from other institutions and through purchase of material.

When the National Air Museum was established, the printed materials were assigned along with the artifacts to the new organization. In 1965, the NASM Library was designated as a Branch of the Smithsonian Institution Libraries but it was not staffed by library personnel until 1972, when, in anticipation of the new museum building, the first Branch librarian was hired and other staff positions were added.

Until 1972, the NASM library consisted mainly of a large donation (The Institute of Aeronautical Sciences Collection), other smaller donations, and the combined curatorial collections described earlier. When library staff members were assigned to the Branch, a systematic approach to the development of the collection began. The existing collection was carefully reviewed so that the backlog could be cataloged and funds were made available for the purchase of additional material. Full-time staff also permitted the organization and initiation of various reference services.

NASM LIBRARY USERS

The NASM Library's primary users are the NASM staff. The principal research programs that require library support are the collection of artifacts and artifact documentation; conservation and restoration; exhibit research and preparation; responding to public inquiries; educational programs such as seminars, lectures, tours, film programs, and the like; special projects such as the Space Telescope History Project; and publication programs ranging from articles by individual research staff to the multi-volume history of aviation now underway. While numbers change regularly, there are at any one time from sixty to sixty-five researchers engaged in historical and technical research in aerospace fields and from twenty-five to thirty staff conducting research in support or ancillary activities.

In addition to the permanent NASM staff, there are varying numbers of visiting scholars, fellows, and interns who are temporarily associated with NASM from periods of a few weeks to a year. These temporary staff members are usually conducting intensive research on a fairly narrow topic in conjunction with a curator. The temporary researchers are constantly changing in numbers and interests and make very active use of library collections and services.

Other Smithsonian personnel whose research requires the use of resources held by the NASM Library have full access to these resources by visiting the library or through intra-library loans.

As a reflection of the great popularity of the aviation and space flight exhibits and of other programs of the Museum, the NASM Library receives substantial use from a wide spectrum of researchers not associated with the Smithsonian Institution. These include historians, biographers, aircraft restorers, staff of other museums, model builders, aeronautical enthusiasts, representatives of the media, educators, and students. An average of approximately 175 such users come to the library to conduct their research each month. Other inquiries from the general public come to the library by telephone or in the mails.

NASM LIBRARY RESOURCES

The *general collection of the NASM Library* is made up of monographs, serials, technical reports, government publications, microforms, and the like. There are approximately 30,000 volumes cataloged as monographs and 7,000 volumes of bound serials. The library holds about 500 serial titles, of which over 300 are received currently. The serial collection includes especially fine holdings of early aeronautical journals. Currently, these holdings are being enhanced by the addition of foreign-language aeronautica.

The subject content of the general collection reflects the varied research programs of NASM. The great majority of the collection is on or directly related to aerospace subjects including aviation history; flight material; air transport; balloons and ballooning; propulsion; military aviation; space history; astronomy; spacecraft design and instrumentation; rocket engine design; lunar, planetary, and terrestrial geology; and remote-sensing. The library collection also includes some basic reference materials to support ancillary NASM programs such as publication preparation and marketing, public affairs, education, new technologies, art design, and fund raising. Additional materials for the latter activities are obtained from other SIL Branch Libraries or through interlibrary loan.

A *reference collection* of about 2,500 volumes consists of some standard general reference works, specialized aerospace reference works, and general aerospace texts that have proven most useful in responding to reference and information inquiries.

Special collection material is housed in the Ramsey Room, which is named after Admiral and Mrs. Dewitt Clinton Ramsey. After her husband's death in 1961, Mrs. Ramsey established a fund at the Museum as a memorial to her husband with some of the funds being used for building the special collections facility. NASM's special collections materials include books, manuscripts, scrapbooks, sheet music, children's literature, newspapers, magazines, and photographs. Collection scope includes balloon history from the earliest times, aviation, science fiction and fantasy, and space exploration, with the book collection numbering about 1,000 volumes. Donations from the Institute of Aeronautical Sciences, now the American Institute of Aeronautics and Astronautics, the William A. M. Burden collection, and the Bella Landauer collection provided a beginning for the collection. Some highlights of the current collection include the Upcott Scrapbook containing accounts of early balloon ascents, the Bella Landauer sheet music collection, an autograph collection, and the papers of Smithsonian Secretary Samuel Pierpont Langley as well as other flight pioneers.

All *Smithsonian Institution Libraries collections* are accessible to NASM research staff through the system-wide online catalog that has been operational since mid-1984. The data base is growing rapidly through new acquisitions and through systematic conversion of old records. At present, it includes a large percentage of all

Smithsonian Institution Libraries holdings and virtually all of the holdings of the NASM Library.

Three commercial *electronic data bases* are available at the NASM Library to support NASM staff research. These data bases are DIALOG, NEXIS, and OCLC.

The NASM Library does not have a vertical file but depends upon the files of the *NASM Records Management Division*. Historically, the two units were administratively separated several years ago. In spite of this separation, the two units are interdependent in terms of functions and services. They work together very closely and share a reading room as well as other facilities.

There are over 30,000 subject files in the Record Management files which consist of millions of items such as photographs, pamphlets, clippings, reports, manuscripts, technical manuals, drawings, and articles covering all aspects of aerospace development including specific aircraft and spacecraft, biographies of important personalities, and significant events. Photographic prints and negatives alone number over one million. The files are fully accessible to library staff members who draw on them heavily in providing research services, and the library staff regularly contribute material to the files.

The *NASM Library staff* consists of two full-time and one part-time professional librarians, one technical information specialist, two library technicians, and one secretary. Staff members perform mainly public-service-oriented functions as described below. In addition, they participate in collection development and in performing administrative tasks such as training, management, supervision, and local as well as Smithsonian-wide planning and coordination. In order to enhance services further, the library has from fifteen to twenty volunteers assisting in various services and projects. Each donates time and talents, ranging from one-half day to three days each week.

NASM LIBRARY SERVICES

Reference and Information Services are provided by means of traditional manual literature as well as machine-assisted searches. Full services are available to Smithsonian staff but because of the limited size of the library staff and the cost of machine-assisted searches, the library cannot provide in-depth research or machine-assisted searches for non-Smithsonian users. The library staff provides guidance to help these users locate and use resources to conduct their own research.

In providing reference and information services, library staff members search pertinent material among the resources mentioned above and, as appropriate, they also consult experts within or outside the Smithsonian Institution. For Smithsonian staff, the research is exhaustive. For non-Smithsonian users, when the answer cannot be provided after a reasonable search, the user is provided with leads for further searching.

The library staff provides an *indexing service* for NASM staff and for other users. Forty-three serials not indexed by other indexing services are reviewed for articles of interest to NASM aeronautics research staff. The index compiled by NASM Library from 1973 through 1982 was published by G.K. Hall in 1983 as *The Aerospace Periodical Index, 1973-1982.* The index since 1982, which has been modified in scope, content, and format, at present exists only on line. Eventually, a cumulation will be considered for publication.

Circulation and interlibrary loan services are available at the NASM Library. The general collection material is available for loan to Smithsonian staff and for inter-library loan, but the Reference and Special Collections materials do not circulate except under special conditions. Non-Smithsonian users may borrow NASM Library material by requesting interlibrary loan through their organization, institution, or public libraries.

NASM Library staff maintains *serials control* of titles held at the branch, both currently received titles and those no longer received but retained. Currently received issues are still being entered manually while the retrospective record of earlier holdings is being converted to an automated system.

NASM Library participates actively in the Smithsonian Institution Libraries *conservation and preservation* programs. Completed volumes of serials, paperback monographs, and volumes in need of rebinding or repair are forwarded on a monthly basis to the central Binding Section for processing to commercial binders. Material from the Special Collection is forwarded, also on a monthly basis, to the Smithsonian Institution Libraries Conservation Laboratory, where conservation experts take appropriate conservation measures.

CONCLUSION

From humble beginnings as the personal collections of Smithsonian research staff, the NASM Library has evolved into a first-rate research library with resources and services that have been developed to serve the specialized requirements of NASM researchers. The research community outside the Smithsonian Institution, world-wide, also has access to its resources and services. While the level of achievement is notable in view of the relatively brief life of the Branch, the NASM Library faces, with enthusiasm, new challenges of continuing development and growth. The goal of NASM is to become the world's foremost historical research center on aerospace subjects and the continued development of the NASM Library is very much a means for the achievement of that goal.

STATISTICS

Name of Museum	National Air and Space Museum
Date Founded	1946
Name of Library	National Air and Space Museum Branch Library, Smithsonian Institution Libraries
Telephone Number	202/357-3133
Name of Library Director	Frank A. Pietropaoli, Branch Chief
Library Collection Size	
Number of Monographs	30,000
Number of Bound Journals	7,000
Number of Microform Journal Volumes	400(reels); 500(fiche)*
Main Subjects Collected	History of Aviation, Aviation Technology, Space Exploration, Space Science, Planetary Studies

*Has over 100,000 additional fiches consisting of NASA documents.

Staff Size	2 Full-time Professionals 1 Part-time Professional 4 Non-professionals
Online searching done	Yes
Interlibrary loans made	Yes
Names of networks affiliated with	FEDLINK, OCLC

National Museum of American History Branch Library

Rhoda S. Ratner

ABSTRACT. The National Museum of American History Branch Library, a part of the Smithsonian Institution Libraries system, acquires and maintains a dynamic collection in support of the research pursued in this museum. Subject concentration is in the history of science and technology and American history, with emphasis on material culture and on documenting and servicing the museum's collections. Special collections include trade literature and exposition materials. The library collection is developed and viewed as a national resource, with access available to all scholars in need of its materials.

The National Museum of American History (NMAH) Library, a branch of the Smithsonian Institution Libraries (SIL) system, supports the research, exhibition and public programs of the Smithsonian Institution as well as the specialized interests of the staff of the National Museum of American History and members of the public with an interest in American history and in the history of science and technology.

This library currently houses a collection of approximately 165,000 volumes of books and bound journals on engineering, transportation, military history, science, applied science, decorative arts, and domestic and community life in addition to American history and the history of science and technology. There are over 500 titles on microform including city directories for selected cities from the eighteenth and nineteenth centuries and United States Patent Specifications from the earliest records of the United States Patent Office through 1906.

Rhoda S. Ratner is Chief Librarian, National Museum of American History Branch Library, Smithsonian Institution Libraries, Washington, DC 20560. She has a BA and an MLS degree from the University of Maryland.

BRIEF HISTORY

The NMAH Branch Library had its formal beginning as an organizational entity in 1958 in the old Arts and Industries Building of the Smithsonian. At that time, the object collections which were the basis for the establishment of this museum were housed there. The library collection was a combination of books from that building as well as the sectional libraries of the departments of civil history, military history, science and technology, and arts and manufacturing. The mission was to supply source materials on the historical and technological development of this country.

At that time, a pattern was established which has endured to the present. Some of the divisions preferred to keep within their own areas those books of active interest to them. Over the years, most of these collections have been cataloged by the Smithsonian Institution Libraries with the books circulated on indefinite loan to the curators.

In today's practice, each curatorial division within this museum has the equivalent of a sectional library with the books charged out to the various researchers within the division. Responsibility for the materials lies with the borrower, and the books are made available to other patrons as the need arises.

In the early years of operation, the library reflected the staff's narrowly defined research interests. As the staff developed broader interests and needs changed, special efforts were made to make the facilities more available and useful to outside scholars. By 1971, the NMAH Branch Library committee recommended that the primary aims of collection policy should be: (1) to provide materials for documenting and servicing the collections and the public contacts they generate; (2) to support curators in their research efforts; and (3) to provide a facility which can serve as a center for scholars in fields of interest to the museum. The current mission of the NMAH Branch has changed little in the intervening years.

The Smithsonian Institution Libraries system maintains several branches throughout the Institution's museums, with each Branch Library the responsibility of a Chief Librarian. Technical support in the form of acquisitions, cataloging, and requisitioning of supplies and equipment are provided centrally. Within the host museum, the Branch Library is administered in close coordination with the operation of the museum without being administratively accountable to the museum.

In the National Museum of American History, for example, there is a library liaison committee composed of representatives from the Department of Social and National History, the Department of History of Science and Technology, the Office of the Director, and the Chief Librarian. This committee meets twice yearly to consider solutions to problems and plans for changes in service as they are required.

The Smithsonian Institution Libraries are organized into three divisions—Research Services, Bibliographic Services, and Collection Development. All Branch Libraries are accountable to both the Research Services and Collection Development Divisions and are further organized into units. The National Museum of American History Branch Library is part of the History and Art Unit, and the four Chief Librarians within this unit report through one unit head.

The staff of the National Museum of American History Branch Library consists of a Chief Librarian, a senior Reference Librarian, three full-time library technicians, and two part-time technicians. The staff is responsible for a full range of reference services and collection development, maintaining our own serial and binding records, requesting and processing all interlibrary loan requests, and searching data bases. In common with all other parts of the Smithsonian Institution, we welcome the assistance of volunteers and are most likely to utilize their help in accomplishing special projects.

SERVICES AND AVAILABILITY

The library is located in a non-public part of the building, and the hours of operation are 8:45 am to 5:15 pm, Monday to Friday, for Smithsonian staff and 10:00 am to 5:15 pm, Monday to Friday, for the public by appointment. The library serves the staff of the Smithsonian Institution, government agencies, professional associations, scholars and the general public, including students and collectors needing access to specialized resources.

The public, including all patrons from outside the Institution, may borrow materials through the interlibrary loan system available to them in public, special, college and university libraries. A photocopying machine is located in the library, and other photo reproduction services for slides and photographs are available through the Smithsonian Office of Printing and Photographic Services. Commercial data base searches are provided only for staff members.

CATALOGS

The Smithsonian Institution Libraries recently began to implement its automated library system which will eventually include fully automated modules for a totally integrated online system. We are now using it for cataloging, acquisitions, and our online catalog. This online catalog includes all materials cataloged for the Smithsonian Institution Libraries since 1965. Prior to that time, materials were cataloged in Dewey. A retrospective conversion project of the older materials is in progress, and these records will be added to the data base as they become available.

There are terminals in all of the Branch Libraries for use by staff and researchers. The system (designed by Geac) is user-friendly, and the switch from a COM catalog to the online system was met with much enthusiasm. In addition to the online catalog, there is a COM 'Smithsonian Institution Libraries List of Serials' which includes records of all Smithsonian serials and is updated every six months. Most serial titles are also found on the online catalog.

For materials cataloged prior to 1965 and not yet converted to machine-readable format, it is necessary to consult the card catalog in the branch library.

The Smithsonian Institution Libraries has extensive collections of printed book catalogs of other libraries in addition to its online access to the bibliographic information available from OCLC and RLIN. We also share in the interlibrary loan system available through OCLC.

RESEARCH SUPPORT

Research in the National Museum of American History is devoted to collecting, preserving, and interpreting artifacts with the special mission of contributing to the cultural, political, and economic history of the United States through research that derives its evidence principally from these material artifacts. As a national museum, the natural focus is on the history of the United States of America, including its roots and connections with other cultures.

At the broadest levels, the collections are concerned with the history of science and technology and their impact on the American scene, and with the everyday life of Americans, including the material aspects of the home and workplace, traditional folk arts and popular culture, enrichment through the visual arts and music, and the political history of the United States.

The library collections are strongest at the research-program level, including materials necessary to identify objects and place them in an historical context. Library materials are used for both research on exhibitions and may also be included as objects in exhibitions. For this reason and for the duplication of illustrations, every effort is made to retain the hard copy. As a national museum, there is also a general level of documentary texts available to answer outside inquiries peripheral to the primary research areas.

The curatorial divisions of the National Museum of American History are as follows: Department of the History of Science and Technology—Divisions of Armed Forces History, Electricity and Modern Physics, Extractive Industries, Mathematics, Mechanical and Civil Engineering, Mechanisms, Medical Sciences, Physical Sciences, Transportation; Department of Social and Cultural History—Divisions of Ceramics and Glass, Community Life, Costume, Domestic Life, Graphic Arts, Musical Instruments, Photographic History, Textiles; the National Numismatic Collection; the National Philatelic Collection; and the Department of Public Programs.

COLLECTION DESCRIPTION

Library materials are added to support current research interests, to enhance the general collection beyond the scope of a single division or curator, and to continue coverage of interest not currently under research but primary to the goals of the museum. Books dealing with objects and artifacts—material culture, steam engines, costume, manufacturing, philately, etc.—are collected as exhaustively as is practical. Books dealing with the impact of these things on society are also of importance to this collection.

Every effort is made to collect materials contemporary with the event or object being researched. These document the current knowledge or the product of knowledge of the time in question. A special emphasis of this museum's research is tracing change in a process or in a theory. Succeeding editions of a single work are therefore collected by the NMAH Branch.

The primary language of the collection is English, with many materials in Western European languages. In some research areas, such as numismatics and philately, where the artifact collections are international, supporting documentation is collected in all necessary languages.

Geographically, the library collection of the National Museum of American History Branch Library provides an important national resource for the full interpretation of American history and aspects of non-American history that have influenced the cultural and technological development of the United States.

The chronological period of most general concern is from the eighteenth to the early twentieth centuries. However, each division defines its period limitation in relation to its research areas.

Monographs and journals form the basis for the library collections, but the single type of material most needed by every division is the trade catalog. Microforms are collected to complete or to replace runs of journals, and to include unique materials which may not be available in any other form. There are no prints, slides or photographs in this branch library.

TRADE LITERATURE

Trade catalogs constitute a primary source of information for documenting the timing, function, value, and place in society of museum objects. Since the early days of this collection, this material has been a major component, with strong efforts being made continually to enhance its value to researchers. Acquisitions are made by gift, exchange, and purchase, and we have been the recipients of several large collections—Columbia University was a major donor, and additional gifts have come from the Baker Business Library of Harvard University and the University of California Library at Davis. Recently, we were offered a collection from the Center for Research Libraries, and that transfer is in process now.

The size of the trade-literature collection has been estimated at approximately 250,000 pieces, and may be the largest collection of its type in this country. The concentration is on nineteenth- and early twentieth-century catalogs, with a full range of subject/object coverage. Purchased catalogs vary widely in price and come to us through dealer catalogs or are discovered by museum curators. Occasionally, a catalog is considered of such special value or rarity that it is transferred to the SIL Special Collections Branch Library for rare-book treatment and storage.

Because of the size of the collection and limitations on our staff, various means of bibliographic control have been employed in the past. In the early days of the branch, the catalogs were classified in-house using the Dewey Decimal system. These numbers were prefixed with a 'C' and intershelved with the regular collection. Approximately 15,000 catalogs were classified in this manner. In order to provide access to the bulk of the collection while these items were being cataloged, the rest of the trade literature was arranged alphabetically by company name, with no written documentation on these holdings. Researchers had access to the open shelves and could search as long as they had company names as leads. The major part of the collection is still arranged in this manner.

In 1982, the decision was made to begin processing the catalogs using an abbreviated form of the MARC record, storing the data in the Smithsonian Institution's computer. The following fields are utilized: company name; title; special collection; place of publication; date of publication; language; subject; holding library; serial status; accession number; and whether it is a microform. Unbound catalogs are placed in acid-free envelopes and all catalogs are shelved by accession number. Access is through a microfiche listing arranged by the company name and subject. The goal is to merge this data with our online catalog so that the trade literature documentation will be available in the same mode as the rest of the Smithsonian Institution Libraries collections.

Of all the decisions necessary before this plan was adopted, the most difficult was the assignment of a single Library of Congress Subject Heading to each catalog. For those companies advertising a broad range of objects, the most encompassing subject is selected, with the consequence that research data is 'lost' on some objects. The justification for the decision is on several levels. From past experience, we knew that most researchers using trade catalogs came with the company names of interest to them. An alternate approach was through the subject, where the researcher could find a listing of all the entries under that subject. If both of these methods failed, our back files of the *Thomas Register* could be searched by subject for company names. While not an ideal method, the single-subject entry is working for us.

Since the beginning of this program in 1982, over 20,000 pieces have been processed through the efforts of one staff person and several volunteers. The abbreviated MARC format and the single-subject designation enable volunteers to participate with a minimum of training. The Cooper-Hewitt Branch Library (design) and the

Horticulture Branch Library (seed catalogues, garden statuary and furniture) also collect trade literature, although the largest number of the catalogs which have been processed in this format have been assigned to the NMAH Branch Library.

The project continues along with a special study now in process to determine the best arrangement for the large number of catalogs currently stored off the Mall which have never been arranged or processed. This study is also designed to investigate available means for preserving the data in these catalogs before they disintegrate beyond the point of preservation. Since our goal is to retain the hard copy whenever possible, access through electronic means would be most desirable, with handling of the hard copy only when absolutely necessary. We value this collection as a major resource for scholars in many disciplines and will seek to preserve it for the future by whatever means possible.

EXPOSITION MATERIALS

Another special collection encompasses exposition materials published by or about international exhibitions or 'world's fairs'. Approximately 1,500 pieces, dating from the 'Crystal Palace' Exhibition (London, 1851) to the present, comprise this collection. The bulk of the collection is cataloged in the Dewey Decimal system with the broad '507' classification followed by the location code. These were processed at a time when full cataloging was not possible and therefore there is no shelf identification beyond the geographic location and date available for specific items. In practice, this means that all materials from the Chicago fair of 1893 are shelved together, and the researcher is required to search within that particular area for the actual title needed. Again, as with the trade catalogs, this is not an ideal solution, but the researchers who come to use this collection are not novices, and the additional opportunity for the serendipitous 'find' often makes the extended search all the more worthwhile.

CONCLUSION

The collections of the National Museum of American History Branch Library provide a sound base for the research conducted here. Changing research emphases are reflected in our collection policy, resulting in a constant state of evolution and development. We are proud that this library has indeed become a national resource for scholars.

STATISTICS

Name of Museum	National Museum of American History
Date Founded	1958
Name of Library	NMAH Branch Library, Smithsonian Institution Libraries
Telephone Number	202/357-2414
Name of Library Director	Rhoda S. Ratner, Chief Librarian
Library Collection Size	
Number of Monographs	120,000 est.
Number of Bound Journals	45,000 est.
Number of Microform Journal Volumes	500 total titles in microform & microfiche
Main Subjects Collected	History of Science, History of Technology, Material Culture, American History, Decorative Arts
Staff Size	2 Professionals 5 Non-professionals
Online searching done	Yes (for SI staff only)
Interlibrary loans made	Yes
Names of networks affiliated with	OCLC, RLIN

Natural History Museum of Los Angeles County Research Library

Katharine E. S. Donahue

ABSTRACT. The Research Library of the Natural History Museum of Los Angeles County has been an integral part of the Museum since it opened in 1913. The Library collections have grown in support of the biological and historical collections. Although open to the public, the Library is primarily a research facility for the curatorial staff and other users of the collections.

MUSEUM DEVELOPMENT

On November 5 and 6, 1913, Los Angeles celebrated in Exposition Park the arrival of two vital resources: water from the distant Owens Valley and the Museum of History, Science and Art. The much needed water would support the growing population which in its turn would support the Museum.

The Museum was a long pursued goal which finally came into being on February 7, 1910, when a contract was made between the County of Los Angeles and the Historical Society of Southern California, the Fine Arts League, the Southern Division of the Cooper Ornithological Club and the Southern California Academy of Sciences. These organizations were to acquire, conserve and exhibit the collections. The County would take care of expenses, maintenance, and operation of the Museum. This cooperation between the County and the private sector has continued to this day.

In 1965 the decision was made to move the art collections to a new facility in Hancock Park. Since that time, the Los Angeles County Museum of Art has operated as a separate institution. Ignoring the history component of the Museum in Exposition Park

---

Katharine E. S. Donahue is Museum Librarian, Research Library, Natural History Museum, Los Angeles County, 900 Exposition Blvd., Los Angeles, CA 90007. She received BA and MLS degrees from the University of California at Los Angeles.

(the original and present site of the Museum), the institution was renamed the Los Angeles County Museum of Natural History, informally called the Natural History Museum of Los Angeles County to avoid confusion with the Art Museum.

The space freed by the departure of Art was used to expand the natural science and history exhibits and to increase storage space for ever growing collections.

In 1977, through the generosity of George C. Page, the Natural History Museum built a satellite museum on the site of the La Brea Tar Pits, called the George C. Page Museum of La Brea Discoveries. This museum, devoted entirely to the study and interpretation of the Pleistocene fossils found in the tar pits, provides the public with information about this incomparable site, and allows scientists to study the wealth of material excavated from the tar.

In the 72 years since its dedication the physical size and plant of the Museum have both grown consistently. The physical plant of the main Museum now contains over 380,000 square feet, while the satellite Page Museum occupies 50,000 square feet. The various systematic collections (consisting of fossil and recent organisms) now contain over 10,000,000 specimens; the ethnographic collections contain over 225,000 artifacts; the historical collections have over 1,500,000 objects and artifacts. There are 38 curators who care for, research and interpret these collections. The curatorial staff are expected to do research and publish the results in scholarly journals; this research may or may not be related to the collection responsibility of the curator.

In addition to the Curatorial Divisions of the Museum, there is a Public Programs Division, whose staff includes publications editors, exhibit designers, model makers, taxidermists, graphic artists, education specialists, printers, and photographers. This cadre of people works closely with the curatorial staff in the creation of education programs and exhibits.

LIBRARY HISTORY

The Research Library has been an integral part of the Museum from the very beginning; it is obvious that the need for a library was recognized early and plans made for it.

The first library collections came into the Museum from the Historical Society of Southern California and the Cooper Ornithological Club. The Southern California Academy of Sciences, as well as staff members, also donated books and materials in the Museum's early years. (The Library still receives exchange material from the *Bulletin of the Southern California Academy of Sciences*; we also receive material in exchange for the *Journal of the Lepidopterists' Society*.)

In the early 1920s the Los Angeles County Public Library designated the Library as depository branch; they assigned a librarian to be in charge and deposited a variety of scientific works in the branch. This proved to be a short-lived arrangement, perhaps because it is inimical to public library spirit to have material be non-circulating, and in 1925 it was dropped as a branch. The Museum assumed full responsibility for the Library and the librarian was appointed Museum Librarian. Since that date, the Library has always been administered by a professional librarian.

Administratively, the Library functions as a Division of the Museum, with the Museum Librarian reporting to the Museum Director. The Museum Librarian is considered part of the management group of the Museum and regularly participates in management meetings concerned with information and the implementation of policy. The Museum Librarian is responsible for creating policies and programs and overseeing them for the Library; she is not however involved with making general policy for the Museum.

The Library staff consists of 3 full-time professional librarians, two of whom have second master's degrees in appropriate subject areas; 1 full-time library assistant; 1 part-time library assistant; and 2 part-time student workers. All the professional staff are involved in a variety of activities and share the responsibilities of public service to the staff and the public. Other activities such as cataloguing are differentiated by the subject specialization of the librarians. Interlibrary loan is arbitrarily assigned and the responsibility for all the multifarious activities connected with serials is divided among several staff members. There is a recognized need among the librarians for a serials librarian.

LIBRARY COLLECTIONS

The Library collection contains more than 250,000 items consisting of monographs, serials, maps and charts, iconographic materials, manuscript and archival materials, photographs, microforms and newspapers. The Library contains over 90,000 book and journal volumes. There are 3,315 serial titles held, of which 1200 titles are currently being received; two-thirds of these are received through exchange of the Museum's scientific publication, *Contributions in Science*.

The subject strengths include New World anthropology, and archaeology, botany, ecology, entomology, ichthyology, herpetology, invertebrate zoology (including malacology) invertebrate and vertebrate paleontology, mammalogy, ornithology, and mineralogy, as well as industrial technology (with an emphasis on transportation), California history, and the history of the trans-Mississippi West and Mexico.

The composition of the Library collection is dictated by the disciplines it serves, the type of research performed by the curators, and by the museum framework. For systematists, because of international codes and agreements, access to old literature (anything published after the 1750s when Linnaeus published his two seminal works on classification) is as important as acquisition of the current literature. For libraries supporting systematic collections, this means a library collection with eighteenth, nineteenth, and twentieth century books and journals. This kind of collection presents problems concerning access and, perhaps more overriding for the nineteenth and twentieth century materials, problems of preservation so that they may be used. Finally, because the material is housed in a museum, it may be looked at for its artifactual value. Thus, material whose content is no longer of value may be saved as an artifact of a time past.

LIBRARY SERVICES

In light of this framework, the Library staff cares for and makes accessible the resources of the collection. The primary users of the collection are the curatorial staff to whom material, except certain rare material, circulates on an unlimited basis. This practice has led to the formation of "sectional libraries". These sectional libraries form and develop over a number of years. The material, consisting of books, individual volumes of interdisciplinary journals, and complete runs of specialized journals, is formally charged to a curator; it is then usually housed in close proximity to the specimen collections in an area designated as the library. They are often arranged by subject and accessed through memory; no sectional card catalogues exist. The main catalogue in the Library contains all the cataloguing, and the circulation file, arranged by call number, provides locations. These sectional libraries exist as specialized support collections and remain intact as long as the curators have room for them, use for them, and are willing to be responsible for their security. Although these libraries serve the curators and visiting scientists of a section, they make service to other curators and the public more time consuming. Despite this, material is routinely retrieved for staff and public. The material does not circulate to the public or any non-staff member with one exception. The Library has reciprocal borrowing privileges with the Hancock Library of Biology and Oceanography, located on the campus of the University of Southern California (USC) directly across the street from the Museum. The Hancock Library is rich in old and important systematic literature and the curators use it frequently. The Museum Library in turn loans materials to Hancock faculty and graduate students.

The Library is also used on-site by scientists and scholars through a formal system of appointments. The Library also makes its collection available through interlibrary loan; we are net loaners of materials and borrow and loan through OCLC. Much of the journal material is photocopied and then sent out; a fee is charged.

The Library maintains a divided card catalogue: author-title and subject. Until 1979, the Library purchased Library of Congress cards. In addition to buying cards for books, the Library also had standing orders for analytic cards for monographic series, such as the *Professional Papers of the U.S. Geological Survey.* In-house analytics for certain monographic series were also created as well as analytics for pertinent articles in various journals not indexed elsewhere. The catalogue is therefore a helpful, although uneven, index to the material in the collection.

In July of 1979 the Library went on-line with OCLC, which now supplies our cards. (Although we are buying MARC tapes, we foresee no early opportunity to use them in an on-line catalogue. The Library has considered a COM catalogue.) We have continued to use and create analytics for the appropriate monographs and articles.

The Library provides access to serial titles through the University of Southern California's union list of serials, in which our holdings appear with our location code. USC provides us with a complete union list as well as a listing that includes just our titles. We have agreed to make our resources available to USC students on a reference basis only. Copying services are available.

In conjunction with the main card catalog, the Library staff uses a variety of published indexes to access the collection and make information available to the staff. Because of the diversity of subjects in the Library and the Museum, a variety of indexes are required. One of the most frequently consulted indexes is *Zoological Record* (worldwide coverage of literature) which began in 1864 and is still being published; 1978-1980 are now available through DIALOG. The *Bibliography and Index of Geology* is used to access the earth sciences literature. It began in 1933, and until 1969 excluded North America; its coverage is now worldwide. GEOREF (DIALOG: geology, mineralogy, paleontology, etc.), covers North American material 1929-present and worldwide 1967-present. The *Bibliography of Fossil Vertebrates* (worldwide coverage), has been variously published since 1928 and is a valuable resource on fossil vertebrates. These indexes all provide a variety of access points, author, subject, etc., for the user.

Another type of index generally required by researchers working with systematic biological collections is an index of names and a genera index. Every species of organism has two names, genus and species, (*Homo sapiens, Felis domesticus*) and these specialized indexes, often arranged by genera, provide reference to the original description (where the organism is formally described and named for the first time) and where it was published. Two such indexes are *Nomenclator Zoologicus* (coverage 1758-1965) by Sheffield Neave for animals and *Index Kewensis* (coverage 1758-1970) for plants. Each discipline has its own specialized catalogues and indexes to the flora and fauna within its purview.

The Library performs a variety of services for its patrons and constituents. The Library staff provides quick reference on a variety of topics, from the address of a museum in Kenya to "What's the largest island in the world?" (Greenland). The Library staff assists Museum staff in the use of the Library and uses interlibrary loan through OCLC to acquire material not owned. Formal and informal relationships among the librarians of the major natural history museums sometimes facilitates the loan of rare materials.

Because the shelving areas are closed to the public, we page all material for public patrons, whether it be in the main library or in a sectional library. We use OCLC as a reference tool when it is appropriate, and use the new M300 Workstation from OCLC when we search a variety of DIALOG databases.

Because most of the professional staff are in one building, the Library is able to maintain extensive serial routing lists. A monthly accessions list, purchased from OCLC, is also circulated to all staff. Staff are notified when a book in their field is acquired. In general staff are notified of material which may be relevant to their work or field of study.

At times the Library is involved in assisting a curator in selecting artifacts for an exhibit. This may mean finding an appropriate illustration, such as an eighteenth century engraving of a specific animal or a nineteenth century lithograph of a whaling scene. In the main it means the librarians must know their collection in both content and form.

The Library has been cooperating actively with other libraries and institutions with similar collections. The Museum Librarian and the Director of the Hancock Library submitted a joint proposal for retrospective conversion to the Department of Education in 1984.

Five major natural history museums (the Academy of Natural Sciences of Philadelphia, the American Museum of Natural History, New York, the California Academy of Sciences, San Francisco, the Field Museum of Natural History, Chicago, and the Natural History Museum, Los Angeles) formed a group in 1984 called the Associated Natural Science Institutions. It is designed to promote cooperation and to facilitate understanding of the goals and purposes of natural history institutions. The libraries of these institutions have been cooperating actively and have submitted two cooperative grant proposals.

In addition to serving our present staff and patrons, the Research Library is building and preserving an invaluable resource for future curators and researchers with different interests and unforeseen needs. This means that the Librarians not only spend time and money on the conservation and preservation of the collections, but they also attempt to build broad general reference collections in anticipation of scientific and historical questions not yet asked.

STATISTICS

Name of Museum	Natural History Museum of Los Angeles County
Date Founded	1913
Name of Library	Research Library
Telephone Number	213/744-3387
Name of Library Director	Katharine E. S. Donahue, Museum Librarian

Library Collection Size	
Number of Monographs	Approx. 40,000
Number of Bound Journals	Approx. 50,000
Main Subjects Collected	Biology, Paleontology, Earth Sciences, Ethnology
Staff Size	3 Professionals 1.5 Non-professionals
Online searching done	Yes
Interlibrary loans made	Yes
Names of networks affiliated with	OCLC, CLASS

John G. Shedd Aquarium Library

Jan Powers

ABSTRACT. Zoo and aquarium libraries are developing as special libraries. The John G. Shedd Aquarium Library is described. Its collection and services are discussed.

INTRODUCTION

Zoological parks and aquariums have existed for thousands of years. The first aquarists who kept fish in captivity were the Sumerians over 4,500 years ago.[1] The first zoos and aquariums were established in the United States during the 1870's. At that time the goal of keeping animals in captivity was ". . . to house as diverse a collection of strange and fascinating creatures as their operators could find."[2] The goal behind keeping animals in captivity has changed as society has changed. Zoos and aquariums have moved beyond the stage of merely exhibiting animals for the curious public. Although public education and recreation are still important services, more recently zoos and aquariums have increased their commitment to animal research and conservation. As part of this commitment zoos and aquariums have established libraries. Zoo and aquarium libraries are unique in that they are responsible for live animals. John G. Shedd Aquarium Library is a resource for information on how to provide and promote aquatic animal husbandry.

HISTORY OF JOHN G. SHEDD AQUARIUM

The world's largest indoor aquarium was a gift to the people of Chicago from John Graves Shedd, president and chairman of the board of Marshall Field and Company.[3] His choice to build an

---

Jan Powers is the Librarian at John G. Shedd Aquarium, 1200 South Lake Shore Drive, Chicago, IL 60605. She has a BA in Political Science and Speech from Indiana University-Bloomington and the MALS from Rosary College.

aquarium was influenced by the notion that instructive entertainment for large numbers of persons could be provided by displaying aquatic animals.[4] Shedd Aquarium opened in 1930 and was referred to as the "Versailles of aquarium buildings"[5] because of the classic Doric columns complemented with the Baroque style curved forms and aquatic motifs in the marble and tile.

Currently the Aquarium houses over 5000 fishes representing 500 species, invertebrates, amphibians, reptiles, and mammals. A 90,000 gallon Caribbean reef tank, located in the rotunda, is home for over 500 tropical fishes, marine turtles, and sharks. Tributaries feature many popular freshwater home aquarium species. Marine jewels and sea anemones are displayed in permanent exhibits. Exhibits at the Aquarium change as the inhabitants change. And the inhabitants change frequently as new species are continually collected.

HISTORY OF THE AQUARIUM LIBRARY

During the planning and first four decades of operation books and other library materials were purchased by the Aquarium, but were never centrally organized and maintained. The endowment of Helen Shedd Keith for the Aquatic Science Center made it possible to formally establish a Library to provide a repository for scientific and educational use. The Library opened in December 1975. Current library facilities include the Information Resource Center, Aquatic Biology Lab Library, and the Volunteer Library. Staffed by a Librarian, a Library Assistant, and volunteers, the Shedd Aquarium Library provides information and services to qualified users.

COLLECTION

The Library collection supports the animal care-taking, reference, and research activities at the Shedd Aquarium. Holdings include 5000 monographs, 200 periodical titles, zoo and aquarium newsletters, annual reports, slides, data files, and the Species File. The primary subject area of the collection is fish identification, husbandry, conservation, and diseases. Substantial materials are kept on the care and husbandry of aquatic mammals, amphibians, reptiles, and invertebrates. Subject areas represented in the collection include aquaculture, marine biology and ecology, freshwater ecology, oceanography, endangered species, and water quality and pollution. Subject areas in which only a few selections are maintained include sport fishing, diving, underwater archeology, and aquarium design and philosophy. Additionally, the Library collects all levels of materials that relate to the Great Lakes.

The monographs are arranged according to the Library of Congress classification with Library of Congress subject headings. Cross references are used to link scientific and common names of animals and to specify family. The periodicals are uncatalogued and are arranged alphabetically by title.

Our Library has developed its own system of organizing the species of animals. The Species File houses articles, pamphlets, reprints, reports, and ephemeral materials not readily indexed, cataloged, or otherwise classified and organized. The Species File is a synthetic system of organizing animal information for practical use. It is organized to show the evolutionary relationships of species and still provide easy alphabetical access.

With the exception of the invertebrates, the Species File is arranged according to class: Agnatha (jawless fishes); Chondrichthyes (cartilaginous fishes); Osteichthyes (bony fishes); Amphibia (amphibians); Reptilia (reptiles); Aves (birds); and Mammalia (mammals). The invertebrates are arranged according to phylum: Arthropoda, Ctenophora, Echinodermata, Mollusca, etc. Within each class or phylum, arrangement is alphabetical by family. Within each family genus species is arranged alphabetically. Access cards are provided phylogenetically so that any taxonomic term will retrieve the file. Indexes are by United States common name(s) and genus species to family and class or phylum. Cross references are made: 1. from general animal to specific species; i.e.,

Whales See Family Balaenidae (Right Whales)
See Family Balaenopteridae (Finback Whales)
See Family Delphinidae (Dolphins)

2. from different common names for the same species; i.e.,

PERUVIAN PENGUIN
also called
HUMBOLDT PENGUIN See Family Spheniscidae

The Species File is kept in hanging file cabinets. This file provides access to information that is lost in more conventional classification systems.

ARCHIVES

The Library has recently taken on the responsibility of the organization and maintenance of the Aquarium Archives. The Archives include newspaper clippings, correspondence, research notes, blueprints, photographs, animal logs, slides, and a few bottled specimens. A few photographs of fish remain that were taken with one of the first four color cameras from the 1930's. At this time there is no adequate indexing system for the Archives and many of the materials are in the process of being restored and housed in new archival boxes.

COMPUTER CAPABILITIES

The Shedd Aquarium Library has an IBM Personal Computer (PC) with dial access capabilities to OCLC and DIALOG. Additionally, the Library uses the PC for circulation statistics, order processing, and acquisitions lists. The subject authority file for the Species File is currently being programmed on the PC. The Library administration is exploring the use of d-BASE III as an indexing tool to coordinate subject descriptors in various library files.

USERS AND SERVICES

The Aquarium Library is primarily a staff resource. Staff use varies, depending on the specific responsibilities of the different departments. For example, the Library provides the Education Department with basic materials for writing classroom and laboratory programs for visiting school groups. The Library staff directs the Education Department Instructors to new materials in aquatic curriculum development.

The information needs of the Fishes Department are very different. The Fishes Department determines the best way to keep the animals healthy. Computer searches using DIALOG are done to determine research in progress on fish diseases, parasites, nutrition, and other areas essential to animal husbandry. The Library staff consults with other zoo and aquarium personnel to check daily log books on the care of a specific animal. Anaesthetics for aquatic animals are researched in the Library to insure proper dosage and use in treating or transporting an animal. Reference materials about new Aquarium inhabitants are gathered by the Library staff and sent to the Aquarist in charge of the animal.

The Exhibits Department requests information relating to graphic design, tank design, and tank construction. Slides and pictures that are maintained in the Library are often checked by Exhibit staff in looking for new ideas.

The Aquarium

Administration may request natural history surveys of specified geographical areas where animal collecting may occur. History of literature and informational analysis surveys are offered to all staff members involved in initiating and developing extensive exhibit projects.

Shedd Aquarium Library, like most other special libraries, provides standard services such as selective dissemination of information, new acquisitions lists, interlibrary loans, and compilation of bibliographies.

Although essentially a Staff Library, other users include Aquarium Education department classes, Aquarium volunteers, Shedd Aquarium Society members, aquatic researchers, interns, and the general public. Non-staff patrons are admitted to the Library by appointment only. The Library staff handles phone and mail questions. Library services range from sending a complimentary bibliography to photocopying articles.

NETWORKING AND COOPERATION

The Aquarium Library uses OCLC and belongs to the Chicago Library System through ILLINET (Illinois Library and Information Network). Interlibrary loan, photoduplication, verification, and reference services are provided. Cataloging is done through OCLC. The Aquarium Library staff is permitted staff access and services to the Library at the Field Museum. The Field Museum Library collection is very comprehensive, and this professional courtesy and cooperative gesture allows the Aquarium Library staff to use a very valuable resource.

More specific to the Aquarium information needs, the Library participates in the AAZPA (American Association of Zoological

Parks and Aquariums) Librarians Special Interest Group. Through AAZPA librarians are working to provide cooperation and resource sharing. A newsletter, *Library News for Zoos and Aquariums*, is edited and produced by Kay Kenyon of the National Zoological Park Library in Washington, D.C. The newsletter provides a forum for librarians to share information and ideas.

FUTURE OF ZOO AND AQUARIUM LIBRARIES

As perceptions change about the world environment and zoos and aquariums become more involved in animal conservation and species survival, more information will be published. The challenge for zoo and aquarium libraries is to maintain that information and provide informational services that enhance the quality and care given to the animals. Zoo and aquarium librarians endeavor to anticipate the information needs, direct the users accordingly, and create a dynamic information exchange.

REFERENCE NOTES

1. Kenyon, Kay. Zoo/Aquarium libraries—a survey. *Special Libraries*. 75(4): 329-334; 1984 October.
2. Sausman, Karen, ed. *Zoological park and aquarium fundamentals*. Wheeling, W. VA.: American Association of Zoological Parks and Aquariums; 1982: xi.
3. Deans, Nora L, ed. *A guide to exhibit animals in the John G. Shedd Aquarium.* Chicago: Shedd Aquarium Society; 1983: History.
4. Chute, Walter H. *Guide to the John G. Shedd Aquarium.* Chicago: John G. Shedd Aquarium; 1944.
5. Roti Roti, Donna. Aquatic Science Center Library, John G. Shedd Aquarium. *Illinois Libraries*. 62(3): 264-267; 1980 March.

STATISTICS

Name of Museum	John G. Shedd Aquarium
Founded	1930
Name of Library	John G. Shedd Aquarium Library
Telephone Number	312/939-2426
Name of Library Director	Jan Powers, Librarian

Library Collection Size	
Number of Monographs	5,000
Number of Bound Journals	200 titles
Other	Species File, Data Files
Main Subjects Collected	Fish Identification, Aquatic Animal Husbandry, Fish Diseases, Aquatic Animal Conservation
Staff Size	1 Professional 1 Non-professional
Online searching done	Yes
Interlibrary loans made	Yes
Names of networks affiliated with	OCLC, Chicago Library System, ILLINET, American Association of Zoological Parks and Aquariums Special Interest Group

Small Science Museum Libraries: Remarks on a Sampling of Data

Robert G. Krupp

ABSTRACT. Consists of an analysis of the data obtained in a recent survey of the libraries of 26 less known museums which emphasize some aspect of science and technology. Data discussed include staff size, collection size, subject specialties, interlibrary loan participation, online search activity and memberships in consortia or networks.

INTRODUCTION

There are probably very few, if any, librarians who are unaware that there exist today various publications listing museums, their locations, their objectives, their collections, and perhaps even a few words about their libraries. Quite often though the science community and its related network of libraries become interested in a particular museum's library for any number of purposes and these "few words" may turn out to be rather inadequate. In fact it may be that unnecessary letters are sent and telephone calls are being made to some museums with unknown library resources, resulting in expensive lost time to both the inquirer and the inquiree. Thus, there appears to be a need for a survey of library services and facilities at some of the more less known science museums.

METHODOLOGY

In order to conduct a limited survey the writer sent out explanatory letters and questionnaires to a group of 51 relatively small, randomly-chosen science museum libraries in the United States and

---

Robert G. Krupp was, until his retirement, Chief of the Science and Technology Research Center of the New York Public Library. His current address is 7 Maple Terrace, Maplewood, NJ 07040. He has a BA (Chemistry) from the University of Buffalo and the MLS degree from Columbia University.

Canada. This obviously is a very tiny sample considering the fact that there are some 6000 museum operations in the United States alone and of these perhaps one-half broadly deal with some form or other of science and technology. Two of the inquiries went to Canadian museums, the remainder to those in the United States and the Virgin Islands. This survey obviously is not intended to collect data for a competitive commercial directory but rather to investigate a rather broad range of data and to evaluate their benefit (if any) for librarians who might wish to consider science museum libraries as information sources (and not necessarily as a last resort).

Initial mailings were made in early October 1984 but the response was so poor that a second mailing was made about five weeks later. The bulk of the responses arrived during the subsequent three weeks but a third mailing (and in some instances even a fourth one) was necessary to elicit even the final group of responses which by February 1985 totalled just over one-half of the initial mailing list. Statistically, this may not be too poor a response but the effort to obtain it seemed a bit disappointing. However, because of the size of the sample, its randomness, and the wild spectrum of the kinds of responses, the value may after all not be so much in the data accumulated but rather in the obvious shades of interest by responders in their own operation and in the purpose of the survey.

Interestingly enough, the 28 responses came from 17 states (primarily the East coast) and one Canadian Province. Non-responses involved eight additional states and one United States possession. Curiously, none of the questionnaires sent to five California locations was returned. Also, astonishingly, two of the mailings were returned by the Postal Service as being undeliverable due to "moved, no forwarding address" and yet these addresses were found in one of the most recent directories covering museums.

There were two rather unusual responses which cannot count in the analyses. One came from a library which was listed in a commercial directory as having "geology" as its subject but turned out to be "geneology" instead. The name of the museum gave no subject clue, hence the amusing malapropism (a not too uncommon one among non-scientists). And, it should be pointed out, this geneology library spent more time and effort corresponding on the matter to make certain what it was that we wanted. In the other case the museum responded as one of science "with no library maintained" but gave no clue at all as to what kind of science it covered or indeed any data at all.

ANALYSIS OF DATA

The data provided in the completed questionnaires was given very often in a slap-dash, disinterested manner; nevertheless all facts and figures are used exactly as supplied despite possible misinterpretations of portions of the questionnaires by the participants. Thus there developed some seemingly weird, or unusual at the very least, summary reports in areas such as collection size elements and staffing. Initial intentions were to expand the sampling considerably but due to the unexpected long delay in receiving the first set of responses, this does not appear to be feasible or productive for the present.

A reproduction of the questionnaire is included as a convenience to the reader when data comparisons are made. (See Figure I.)

Usually there were no problems with the museum name and the name of its affiliated library, when the latter was different. In most instances the telephone number given was that of the museum in general. Obviously the larger the library the more likelihood of a separate telephone number. This was true in only a very few cases. Year of establishment of both the museum and library was requested so that a judgment could be made on how recent the information or acquisitions available might be. The range of founding dates for museums was 1824-1981 but with the bulk in the 1960s and 70s. The range for the libraries was 1824-1982, with the bulk running a few years later than the museums. However, not all respondents provided that data. The geographic location was readily provided but oddly enough, perhaps because it wasn't specifically requested, the ZIP code was omitted in most instances.

The number of professional librarians employed averaged 1.23 but that calculation included 15 with no such staff. The nonprofessional staff, including volunteers, averaged 2.69, which involved nine libraries with no such staff and one operation with 23.

Although up to five of the most important topics covered by the library was permitted, only twenty chose to report at all (and interestingly two were zero). The average number of topics was only three. However, 49 subjects were listed. (See Table I for a key to subjects given for specific libraries.)

The single area where one would think considerable care would be taken in reporting was the collection size data. Of the 15 libraries which reported a monographic collection only six had less than 5000 volumes. Two of the not-so-small collections were indicated as un-

FIGURE I.

QUESTIONNAIRE

```
Museum name:
Library name:
Telephone no.:___/___/____
Year established
  Museum:
  Library:
Location (Street and city):
Name of Head of Library:
Title of Head of Library:
Title of Supervisor of Head of Library:
Library staff size:
  Professionals:
  Others:
Main subjects collected (up to five most important topics):

Collection size
  Monographs (Vols.):
  No. of serial subscriptions:
  No. of periodical volumes (Bound)
                            (Microfilm reels):
                            (Microfiche sheets):
  Phonorecords:
  Films:
  Vertical files (Number of drawers):
  Technical reports and pamphlets:
  Patents:
  Photocopies:
  Theses:
  Art (Framed or unframed and artifacts):
  Other:

Professional staff of museum (Total):
Average number of users daily:
  Staff members (professional and nonprofessionals):
  Public (if allowed entrance):
Interlibrary loans made:
  If yes, frequency:
Scheduled hours open during week (total):
Area of library in square feet (including work areas):

Number of seats for users:
Online searching:
  If yes, number of searches per month:
    Vendors:
Network affiliation:
  If yes, names of networks:
Exhibits
  Permanent:
  Rotating:
    If yes, frequency:
                                   Date__________
```

cataloged and in storage. Fifteen libraries also reported serial subscriptions but here the bulk indicated only relatively small collections (generally under 50 titles). A rather high number of bound volumes were reported by nine libraries, some with as many as 250,000. Of the remaining collection sizes were reported as being rather small. For example, only four libraries reported having patents; the numbers ranging from five to three million. Some seven operations showed considerable interest in art and artifacts, one with 40,000 pieces! However, it is important in all these cases to remember that many definitions of these collection categories are possible, and "standards" are frequently adjusted by staff to reflect the museum's goals and the understanding of nonprofessionals (as good as they might indeed be).

Reported professional staff sizes of the museums ranged from

Table I

Library subject index: Following each specific or general subject are numbers which correspond to the 26 libraries participating in this report

agriculture--21

anthropology--3, 4

archeology--3

astronomy--4, 9, 16

aviation--21

biographies of oil personages--24

botany--4, 9

chemistry--15

communications--21

company histories--24

computer science--16

EARTH SCIENCES-- 1-3

energy--17, 18

engineering--8, 15

environmental science--9, 18

ethnology--3

fire equipment--20

GENERAL SCIENCE-- 4-8, 15

geology--2, 4, 11

graphic design--8

health education--10

history of oil and gas industry--24

history of science--21, 26

horticulture--9

labor history--19

lepidoptera--2

LIFE SCIENCES-- 9-13

mathematics--15, 16

medicine--10

metallurgy--26

metal mining--26

natural history--6, 12

nuclear weapons--17

ornithology--11

paleontology--2

petroleum--24

PHYSICAL SCIENCES-- 14-21

physics--15

pollution--9

science education--6, 16

space and rocketry--14

TECHNOLOGY AND ENGINEERING--8, 15, 22-26

textiles--19

transportation--21, 22, 23, 25, 26

uranium handling--17

waste isolation--17

water power--26

whaling industry--13

zoology--4

over 100 down to one, but the average for 19 museums was 23. The daily use of the library was surprisingly low for both staff (generally ten uses per day) compared to six uses per day by the public (if allowed entrance). Frequently appointments were indicated as necessary.

Only 21 libraries reported on interlibrary loan activity: eleven "no" and ten "yes". In the latter case the number of loans per month was virtually nil except for one case: 1500 per month. Here is a situation where it probably can be assumed that no answer at all indicates no interlibrary loans are available.

The areas of the libraries, including work space, for the 17 which reported, averaged to just over 6,000 square feet, but the range was from 200 to 43,000, with most running under 5,000. The average number of seats available to researchers reported was 23 although (of 18) only five indicated a number over 20.

Over twice as many libraries reported on this subject (23) are not involved in online searching. And those which said "yes", except for a rare one or two cases, usage was indicated as seldom. In most cases no vendors were cited. A similar picture developed for the network affiliation question. Of the 24 respondents, 15 said "no", nine "yes" but of the latter, most provided the names of networks involved.

Permanent exhibit activities was a toss-up for the 20 respondents, but of the 14 who reported on rotating exhibits, eleven said "yes" but the frequency was very low.

What conclusions can be drawn from this almost disparate collection of data? Given here are three broad obvious comments, based not only on the tiny segment of responses but also on the lack of responses.

1. There is a need for an other-than-geographic directory on this kind of library. It should probably be arranged by detailed subjects and/or collection categories such as films, phonorecords, art, manuscripts, scrolls.
2. There is a need for the development of a greater interest in supporting the information community as a whole. In fact, a review of ethical considerations within that community vis-à-vis such support may be vital.
3. There is need for continuing the education of science museum library staffs in areas such as online searching and network affiliation advantages.

Following are brief sketches of specific museums and/or libraries which responded. They are arranged alphabetically under five major subject categories. Table I lists those subjects provided in completed questionnaires. After each subject are numbers representing the 26 reporting libraries.

EARTH SCIENCE

1. *EARTH SCIENCE MUSEUM.* 106 Main Street, Loma, MT 59460. Tel 406/739-4357. No library per se. However, as a member of the Museum Association of Montana, of which six do have libraries, loans are available. Founded 1971. Subject: Earth sciences.
2. *HICKSVILLE GREGORY MUSEUM, L.I. Earth Science Center Library.* Heitz Place, Hicksville, NY 11801. Tel 516/822-7505. Richard Evers, Lbrn., Historian-Asst. Dir. Founded 1973. Library holdings: a few monographs, periodical volumes, and slide sets. Not a working library; available to the staff for the most part. Subjects: Geology, lepidoptera, paleontology.
3. *NEW JERSEY STATE MUSEUM, Archaeology/Ethnology Research Library.* 205 West State Street, Trenton, NJ 08625. Tel 609/292-8594. Karen Flinn, Archaeology/Ethnology Registrar. Founded (date unknown). Library holdings: 700 monographs, 30 serial sub. reports, photocopies. ILL. Subjects: Anthropology, archeology, ethnology.

GENERAL SCIENCE

4. *BUFFALO MUSEUM OF SCIENCE, Research Library.* Humboldt Pkwy, Buffalo, NY 14211. Tel 716/896-5200. Marcia Morrison, Cf Librn. Founded 1861. Library holdings: 18,000 monographs, 620 serial sub., 20,000 bd. vol., microforms, rare bks, mss, Oriental scrolls. ILL. Subjects: Anthropology, astronomy, botany, geology, zoology.
5. *(The) DISCOVERY CENTER.* 321 South Clinton Street, Syracuse, NY 13202. Tel 315/425-9068. Stephen A. Karon, Ex. Dir. Founded 1981. Library holdings: small collection. Essentially for staff use. No ILL. Subject: General science.

6. *MUSEUM OF SCIENCE*. Science Park, Boston, MA 02114-1099. Tel 617/723-2500. Edward D. Pearse, Librn. Founded 1831. Library holdings: 35,000 monographs. Over 200 serial sub. ILL. Subjects: General science, natural history, science education.
7. *MUSEUM OF SCIENCE*. 3280 South Miami Avenue, Miami, FL 33129. Tel 305/854-4247. No librn., all volunteer staff. Founded 1910. Library holdings: 3000 monographs. No ILL. Subject: General science.
8. *ONTARIO SCIENCE CENTRE*. 770 Don Mills Road, Don Mills, Ont., Canada M3C 1T3. Tel 416/429-4100 ext. 235. Jeanne Duperreault, Librn. Founded 1966. Library holdings: 8000 monographs, 100 serial sub., 500 bd. vols., 600 films, 30,000 slides, 10,000 prints and negatives. ILL. Public use by appointment only. Subjects: Engineering, general science, graphic design, technology.

LIFE SCIENCES

9. *FERNBANK SCIENCE CENTER*. 156 Heaton Park Drive NE, Atlanta, GA 30307. Tel 404/378-4311. Mary Larsen, Librn. Founded 1967. Library holdings: 10,400 monographs, 355 serial sub., 6400 bd. vols., microforms. ILL. Subjects: Astronomy, botany, environmental science, horticulture, pollution.
10. *HOUSTON MUSEUM OF MEDICAL SCIENCE*. 1 Hermann Circle Drive, Houston, TX 77030. Tel 713/529-3766. No library per se but there is a collection of 25 films. No ILL. Subjects: Health education, medicine.
11. *NEW JERSEY STATE MUSEUM, Bureau of Science Research Library*. 205 West State Street, Trenton, NJ 08625. Tel 609/292-6330. Shirley S. Albright, Science Registrar. Founded (date unknown). Library holdings: 1700 monographs, 25 serial sub., ca. 5000 bd. vols., 3500 reprints. ILL. Subjects: Geology, ornithology.
12. *PEABODY MUSEUM OF NATURAL HISTORY*, with eleven separate divisional libraries. 170 Whitney Avenue, New Haven, CT 06511. Tel 203/436-0639. C. MacClintock, Librn. Founded (date unknown). Library holdings: 14,700 bd. vol., 105,500 reprints. No ILL. Subject: Natural history.

13. *SAG HARBOR WHALING MUSEUM*, John Jermain Memorial Library. This library houses the museum's collection and is directly across the street but is not related to the museum. Main Street, Sag Harbor, L.I., NY 11963. Tel 516/725-0770. James C. Asher, Library Dir. Founded 1910. Library holdings as related to the museum: log books, whaling references, historical artifacts. ILL. Subjects: Whaling industry.

PHYSICAL SCIENCES

14. *ALABAMA SPACE AND ROCKET CENTER*, Wernher von Braun Library. Tranquility Base, Huntsville, AL 35807. Tel 205/837-3400. Edward O. Buckbee, Librn. Founded 1977. Library holdings: 10,000 monographs, 0 serial sub., 5000 bd. vol., 45 microfilm reels, 500,000 technical reports and pamphlets, 5000 photocopies. No ILL. Subjects: Space science, rocketry.
15. *BRADBURY SCIENCE MUSEUM*, Los Alamos National Laboratory Library. Los Alamos, NM 87545. Tel 505/667-4448. J. Arthur Freed, Hd. Librn. Founded 1943. Library holdings: 150,000 monographs, 8250 (4500 titles) serial sub., 200,000 bd. vols., 350,000 technical reports and pamphlets (most on microfiche), 500 films. ILL. Subjects: Chemistry, engineering, general science, mathematics, physics.
16. *FRANKLIN INSTITUTE SCIENCE MUSEUM.* 20th Street and The Parkway, Philadelphia, PA 19103. Tel 215/448-1227. Charles Wilt, Librn. Founded 1824. Library holdings: 65,000 monographs, 500 serial sub., 250,000 bd. vols (includes microfilms), trade catalogs, 3,000,000 patents. ILL. Subjects: Astronomy, computer science, mathematics, physical sciences, science education.
17. *NATIONAL ATOMIC MUSEUM.* Bldg. 20358, South Wyoming Blvd., Kirtland AFB, NM 87117 (Mailing address: POB 5400, Albuquerque, NM 87115). Tel 505/844-4378. Loretta Helling, Librn./Registrar. Library holdings: 0 monographs, 24 serial sub., 30 microfilm reels, 2000 microfiche, 2000 technical reports and pamphlets, 11,600 photocopies, some art works. No ILL. Subjects: Energy, nuclear weapons, uranium handling, waste isolation.

18. *HANFORD SCIENCE CENTER*, Energy and Environment Library, Box 800, Richland, WA 99352, and Dept. of Energy Reading Room, c/o Hanford Science Center, Fed. Bldg. #157, Richland, WA 99352. Tel 509/376-6374 (library), 509/376-8583 (reading room). Terri Traub, Public Information Spec. Founded 1975 and 1983, respectively. Library holdings (combined): 1350 monographs, 30 serial sub., 42 films, 2550 technical reports and pamphlets. No ILL. Subjects: Energy, environmental science.
19. *MUSEUM OF AMERICAN TEXTILE HISTORY.* 800 Massachusetts Avenue, N. Andover, MA 01845. Tel 617/686-0191. Clare M. Sheridan, Librn. Founded 1960 but opened to public in 1964. Library holdings: 20,000 monographs, 30 serial sub., 1400 bd. vol., 125 microfilm reels, 40,000 prints/photographs, some uncataloged technical reports, pamphlets, photocopies, and theses, about 850 patents. ILL. Subjects: Labor history, textiles.
20. *(The) NATIONAL FIRE MUSEUM, Inc.* 21 Endicott Street, Newton Highlands, MA 02161. Tel 617/527-2724. Contact: Norton D. Clark. No library per se but there is a collection of photos, negatives, and glass plates. No ILL. Subject: Fire equipment.
21. *NATIONAL MUSEUM OF SCIENCE AND TECHNOLOGY.* Ottawa, Canada K1A 0M8. Tel 613/998-4566. Minda A. Bojin, Librn. Founded 1967. Library holdings: 10,600 monographs, 400 serial sub., 3250 bd. vols., 40 microfilm reels, 3000 technical reports and pamphlets. ILL. Subjects: Agriculture, aviation, communications, history of science, transportation.

TECHNOLOGY AND ENGINEERING

22. *NATIONAL MUSEUM OF TRANSPORT.* 3015 Barrett Station Road, St. Louis, MO 63122-3398. Tel 314/965-6885. Dr. John P. Roberts, Secy. Founded 1940. Library in storage but collecting is active. Library holdings: 10,000 monographs, over 100 serial sub. Possible ILL. Subject: Transportation.
23. *OWLS HEAD TRANSPORTATION MUSEUM, Museum Research Facility.* Route 73, Owls Head, ME 04854. Tel 207/

594-9219. James Rockefeller, Jr., Librn. Founded 1976. Library holdings: 3000 monographs, 20 serial sub., 20 films, 500 technical reports and pamphlets, 100 pieces of art. No ILL. Subject: Transportation.

24. *PERMIAN BASIN PETROLEUM MUSEUM, LIBRARY AND HALL OF FAME, The Archives Center and Library.* 1500 Interstate 20 West, Midland, TX 79701. Founded 1975. Mrs. Betty Orbeck, Archivist. Library holdings: 1000 monographs (plus ca. 1200 in storage), 7 serial sub., ca. 25 bd. vols., 63 microfilm reels, 800 phonorecords, 40 films, a large number of vertical files, technical reports and pamphlets, manuscripts, maps, photographs, equipment catalogs. ILL. Subjects: Biographies of oil personages, company histories, history of oil and gas industry, petroleum.
25. *SHORE LINE TROLLEY MUSEUM*, George J. M. Kuhn Memorial Library. 17 River Street, East Haven, CT 06512. John R. Stevens, Librn. Founded ca. 1960. Library holdings: uninventoried and uncataloged. No ILL. Subject: Transportation.
26. *WESTERN MUSEUM OF MINING AND INDUSTRY*. 125 Gleneagle Drive, Colorado Springs, CO 80908. Tel 303/598-8850. Peter M. Molloy, Dir. Founded 1982. Library holdings: 7500 monographs, 6 serial sub., 500 bd. vols., 200 microfilms, 1500 technical reports and pamphlets. No ILL. Subjects: History of science, metallurgy, metal mining, transportation, water power.

SPECIAL PAPER

Source Indexing in Science Journals and Indexing Services: A Survey of Current Practices

Virgil Diodato
Karen Pearson

ABSTRACT. The authors discuss source indexing—indexing data that are published simultaneously with the articles they represent. In a study of 685 science journals it is found that: about one-quarter of the journals employ source indexing, especially in the physical and biological sciences and to a lesser extent in the earth and environmental sciences; this type of indexing information usually appears on the title page of articles, a major exception being Chemical Abstracts Service Registry Numbers, which are found at the end of articles; typically, the indexing information is a symbol assigned from a controlled vocabulary, such as the Universal Decimal Classification or the Physics and Astronomy Classification Scheme; also used a great deal are keywords apparently derived from the titles and/or texts of articles; the author of an article often is the initial provider of the indexing data, and one or more editors sometimes check the author's choices. Most of the twenty-three indexing services surveyed indicate that source indexing is a guide to indexing, for usually it is neither disregarded completely nor blindly accepted by indexers.

Virgil Diodato is Assistant Professor and Karen Pearson is Research Assistant at the University of Wisconsin-Milwaukee School of Library and Information Science, P.O. Box 413, Milwaukee, WI 53201. Diodato holds a BA (Mathematics) from Hunter College, an MLS and MA (Mathematics) from Ball State University, and a PhD (Library and Information Science) from the University of Illinois. Pearson holds a BA (History) from North Park College and will receive an MLS in May 1985 from the University of Wisconsin-Milwaukee.

INTRODUCTION

Source indexing of a journal article is a device that can help a prospective reader decide whether or not to read the article. Source indexing occurs when an issue of a journal simultaneously publishes an article and the indexing information, such as keywords or classification symbols, that describe that article. If *Science & Technology Libraries* had provided indexing for the article you are reading now, you might have seen a list of keywords (perhaps "author indexing, controlled vocabularies, keywords") somewhere near the top of this page. Or there might have been a classification symbol, such as 025.3 from the Universal Decimal Classification, in the upper right hand corner of the page. There are many formats in which the indexing could appear, but whatever the method, you would have been able to add this information to the reading of the title and the abstract to give you a feeling for the pertinence of this article to your interests. This, source indexing and an article's title page abstract, share similar purposes, but indexing takes up less space, can be read more quickly, and can be prepared more readily than the abstract. You, the reader, are not the only potential user of source indexing. The indexing also could be helpful to those who route a journal through selective dissemination of information systems, so that articles are pointed to appropriate readers in an organization. And services that compile indexes of journal literature could use source indexing information as an aid to producing their products.

The purpose of this paper is to determine the state of source indexing in science journals and indexing services. We ask five questions about the source indexing that a reader can find today. (1) In which science fields is one likely to find source indexing? (2) What are the various formats in which one will see source indexing? (3) Are the indexing terms assigned from controlled vocabularies, such as thesauri and subject heading lists? (4) Who decides what the indexing should be for a given article? (5) Do indexing services use the indexing found in journal articles?

THE LITERATURE

SOURCE INDEXING AS A READER'S AID. Source indexing can be as useful and ubiquitous as title page abstracting has been for many years. The formal call for title page abstracting came from the

1949 International Conference on Abstracting, which recommended that an abstract appear with "all original articles" published in a journal.[1] However, it was necessary to wait until the 1960's for the appearance of major source indexing programs. These included programs of the American Institute of Chemical Engineers (AIChE) and the American Institute of Physics (AIP). Explaining in 1961 that a literature search "to obtain information vital for technological advancement is presently a tedious, costly, but necessary task", the AIChE began publishing in its journal, *Chemical Engineering Progress*, "catalog cards" with indexing data.[2] A typical card, one for each article in the journal, was a rectangular area that contained indexing terms, an abstract, and a brief bibliographic citation. A card appeared in the same issue of the journal as its article. As more chemical journals began using them, the cards were placed in many different places: on title pages, on the last pages of articles, or clustered together on one or two special sheets at the beginning or end of the issue. The AIP had a similar program, which began in 1963 when one of its journals, *Applied Physics Letters*, published indexing information with some of its articles. The indexing terms appeared in an "indexing categories" or "content analysis" box on the title page of an article. For the AIP the box was the "first approximation" in improving the "reference retrieval system for research physicists."[3]

SOURCE INDEXING AS AN AID TO INDEXING SERVICES. If the first approximation in improving information retrieval was a modification of journal content, then the second approximation involved modification of indexing tools. Comments from decades ago verify that source indexing information could help avoid the duplication of effort that occurred when both a journal and an indexing service index that journal's articles. The AIChE hoped to avoid the "waste of technical manpower" like the situation in which thirty-eight light-metals journals each were processed by twelve indexing and abstracting services.[4] Therefore, the Engineers Joint Council called source indexing a means "to reduce costs and minimize duplication of intellectual effort. . . ."[5]

EARLY SURVEYS ON THE USE OF SOURCE INDEXING. Surveys on the incidence and use of source indexing occurred especially in the 1960's, and these focused on the engineering literature. In 1961, Herner and Johanningsmeier surveyed 1,915 subscribers to AIChE publications. There were 648 subscribers who said they used the source indexing/abstracting catalog cards in the society's journals. Even though a major purpose of the cards was to help readers maintain their own collections of retrospective articles, the survey found that the cards were especially useful to readers of the current literature:

> A very strong majority, approximately 80%, used them not as tools for setting up files for retrospective searching, but as a means of determining the current contents of A.I.Ch.E. publications.[6]

In 1965, the Engineers Joint Council (EJC) figured that seven percent of the 60,000 engineering articles published that year were accompanied by both indexing terms and abstracts.[7] Two years later, the *Engineering Index* "advised that some use is being made of" indexing terms published in primary journals, but the EJC had "no data" on how other indexing services might have been using these data.[8] Finally, a survey discussed in 1968 found that 68.5 percent of 1,150 respondents (all subscribers to AIChE journals) used the indexing terms that accompanied articles. "The major uses of the index terms were as a 'guide for browsing or reading' . . . and as a vehicle for 'current awareness' . . . ."[9]

RECENT COMMENTS RELATED TO SOURCE INDEXING. Current thoughts about electronic publishing have given new life to the idea of accompanying articles with indexing information. One way for electronic publishing to work is to have authors use word processing computers when writing their articles and then to send the texts of the articles from their computers to editors' computers via telephone, satellite, or other communications lines. When an article is published in an electronic journal, the editor places it in a computer to which editor, author, and readers have access. An author can perform the initial indexing of an article, and the indexing information then can stay with the text of the document for as long as the text itself is stored in someone's computer. Price suggests that more than ever before, the presence of the electronic journal will make indexing information and abstracts marketing tools that encourage or discourage readership. This would "impel both authors and publishers to take a greater interest in the indexing of articles . . . . As a result, author indexing could become as common as author abstracts are now."[10] This concept applies to books as well as to articles. The role of the author in electronic publishing is receiving impetus from the Association of American Publishers'

(AAP) Electronic Publishing Project, which is attempting to standardize the elements of an electronically produced manuscript.[11] The Project reports that one of the parts of such a manuscript could be a "subject term . . . .: A word or phrase describing the content of an item."[12] This data element would be the electronic counterpart of the printed journal issue's source indexing information. Because organizations like the AAP could help standardize the elements that comprise an electronic journal, articles and books with accompanying indexing information could be very common in the future.

METHODOLOGY

To answer the five questions raised in the Introduction, we studied journals that are published in the United States and that are filed under thirteen subject categories in *Ulrich's international periodicals directory* (1984 edition).[13] The categories were selected by subjectively assigning a major Library of Congress class to each of the 110 *Ulrich's* categories and then using only those categories in the Q's. The thirteen categories were: astronomy (QB), biology (QH), chemistry (QD), computers (QA), conservation (QH), earth sciences (QE), energy (QC), environmental sciences (QH), mathematics (QA), meteorology (QC), physics (QC), sciences (comprehensive works) (Q), and water resources (QH). We did not use any entries listed in the abstracting and indexing subsection following each subject category in *Ulrich's*.

A total of 1,945 *Ulrich's* titles met these criteria. These then were matched against the serials holding list at the University of Wisconsin-Milwaukee Golda Meir Library, which held 608 of the 1,945 titles. As Table 1 indicates, we personally examined these 608 titles and weeded out eleven directories, such as the *Datapro Directory of Small Computers*, and fifty-five newsletters (and other publications without signed articles), such as *Groundwater Newsletter.*

We then viewed the remaining 542 journals held at Golda Meir Library to identify which journals contained some form of source indexing. For those that did have source indexing, we recorded the format of the indexing and we sent postcard questionnaires to the journal editors to elicit further information about the source indexing contained in their journals. We asked who provides the indexing and whether or not the indexing information is assigned from a controlled vocabulary.

Table 1

Methodology: Accounting for 1,945 Ulrich's Titles

Category	Number
Ulrich's Titles Personally Examined	608
Examined Titles Ommitted (Directories)	11
Examined Titles Ommitted (Newsletters)	55
Examined Titles Remaining (Journals)	542
Examined Journals Having No Source Indexing	398
Examined Journals Having Source Indexing	144
Source Indexing Journals Responding to a Questionnaire for Further Information	81
Source Indexing Journals not Responding to a Questionnaire for Further Information	63
Ulrich's Titles Unexamined	1,337[a]
Titles Unexamined but Sampled and Sent Questionnaire	323
Unexamined/Sampled Titles Not Responding	125
Unexamined/Sampled Titles Responding to Questionnaire	198
Responding Titles (Directories)	13
Responding Titles (Newsletters)	42
Responding Titles Remaining (Journals)	143
Sampled Journals Having No Source Indexing	125
Sampled Journals Having Source Indexing	18[a]

[a]Eventually we did personally examine and verify that 18 of the sampled journals indeed had source indexing.

After examining the 608 journals held by Golda Meir Library, we had to consider the other 1,337 titles found in *Ulrich's*. We randomly sampled about one-quarter of these titles and sent questionnaires to the editors of 323 titles, asking if the titles were newsletters or directories or journals and if they had source indexing. For those who indicated their journals contained some form of source indexing, we personally verified their responses either by locating the journals in the Milwaukee area or by receiving representative samples of the journals through interlibrary loan. Our rationale for personally verifying positive indications of source indexing was to maintain a consistency in the survey. Since all source indexing from journals held in Golda Meir Library were verified personally, we felt that those randomly sampled journals that indicated use of source indexing also should be verified personally.

We entered information about the journals with source indexing into the SCSS statistical program for tallying data.[14] The information included: (1) the subject coverage of each journal, (2) the format of its indexing, (3) who provided the indexing, and (4) whether the indexing came from a controlled vocabulary.

To find out how, or if, source indexing is used by an indexing service, we sent a questionnaire and a sample of source indexing to each service that *Ulrich's* identified as an indexer of at least one of the journals in which we found source indexing.

It is possible that our methodology has caused us to present a somewhat distorted view of source indexing. There were at least three potential problems. First, we studied only some *Ulrich's* journals. Of 1,945 *Ulrich's* titles in science, we limited ourselves to the 608 titles to which we had immediate library access plus the 198 titles whose editors responded to a questionnaire. So, 1,139 *Ulrich's* entries were not considered in the study. The titles we did study might be representive of a given reader's science collection only insofar as the holdings of the library at the University of Wisconsin-Milwaukee are similar to the reader's holdings. Second, our selection of only some *Ulrich's* subject categories omitted study of fields like engineering and the medical sciences. Third, our convenient operational definition of a journal was imprecise: a publication listed in *Ulrich's* and containing signed articles while not being titled as a newsletter or directory. Some of our judgments as to what was or was not a journal probably were subjective.

FINDINGS

Of 542 *Ulrich's* science journals that we examined personally, 144 contained source indexing. (Eighty-one of the 144 journals provided additional information about their source indexing practices, via responses to a mail questionnaire.) Of 143 other *Ulrich's* science journals that we examined indirectly (via responses to the survey of 323 sampled journals), we verified that eighteen contained source indexing. See Table 1.

WHICH SCIENCE FIELDS USE SOURCE INDEXING? Although most science journals in our study did not contain source indexing, those journals that did were concentrated in physics, chemistry, and, to a lesser extent, in biology, computers, mathematics, and astronomy. As Table 2 indicates, 162 of 685 (23.6 percent) journals included source indexing. Above average (mean) rates of source indexing occurred in physics (40.0 percent), chemistry (38.2 percent), biology (28.7 percent), computers (25.4 percent), and mathematics (24.3 percent). The rate in astronomy (22.2 percent) was very close to the mean. Below average rates were found in the environmental sciences (15.4 percent), meteorology

Table 2

Journals, by *Ulrich's* Subject

Ulrich's Subject Category	*Ulrich's* Titles	Journals Studied[a]	Journals (%) with Source Indexing
Astronomy	35	9	2 (22.2%)
Biology	370	167	48 (28.7%)
Chemistry	145	68	26 (38.2%)
Computers	368	63	16 (25.4%)
Conservation	129	33	0 (0.0%)
Earth Sciences	116	64	5 (7.8%)
Energy	138	19	2 (10.5%)
Environmental Sciences	176	39	6 (15.4%)
Mathematics	110	74	18 (24.3%)
Meteorology	20	8	1 (12.5%)
Physics	165	80	32 (40.0%)
Sciences (Comprehensive)	103	38	4 (10.5%)
Water Resources	70	23	2 (8.7%)
Total	1,945	685	162 (23.6%)

[a]Studied by personal examination and/or by a mail survey.

(12.5 percent), energy (10.5 percent), comprehensive works in the sciences (10.5 percent), water resources (8.7 percent), the earth sciences (7.8 percent), and conservation (in which none of thirty-three journals employed source indexing).

WHAT ARE THE SOURCE INDEXING FORMATS? The typical indexing format was the use of information labeled as "keywords" placed at the bottom or top of the title page of the article. Certainly there were other formats. In fact, the 162 journals noted in the previous paragraph included 172 instances of indexing, because a few journals employed two or three indexing formats. Table 3 shows that the most highly used formats were keywords, Universal Decimal Classification (UDC) numbers,[15] and Physics and Astronomy Classification Scheme (PACS) symbols[16]—each placed on the title page of an article. A few journals used Chemical Abstracts Service (CAS) Registry Numbers,[17] the American Mathematical Society (AMS) classification symbols,[18] or *Computing Reviews* (CR) categories and general terms.[19] Other indexing formats used terms labeled as index words, index terms, subject headings, section headings, abbreviations/trivial names, and OR/MS (*Operations Research/Marketing Science*) classes.[20] All but the Registry Numbers occurred almost always on title pages; the Registry Numbers consistently were found on the final pages of articles.

ARE INDEXING TERMS ASSIGNED FROM CONTROLLED VOCABULARIES? Although the keyword was the single most used type of indexing information, most indexing came from controlled vocabularies. Indexing employed keywords in sixty-nine of 172 cases. (See Table 3.) We assumed that the keywords were the results of derived indexing; that is, whoever did the indexing selected keywords by selecting important words from the article itself. This assumption follows from two facts: (1) our reading of journal instructions to authors, which in no case suggested that keyword indexing came from a controlled vocabulary; (2) our survey mailings to editors, none of whom said that keywords were assigned from a controlled vocabulary.

Table 4 shows that at on at least ninety-one of 172 occasions, indexing information was assigned, that is, it came from controlled vocabularies, such as the UDC and PACS lists. We do not know about twenty-eight of the 172 cases, because of absent or incomplete responses to our mail questionnaires.

WHO DOES THE INDEXING? In ninety-one of 172 cases examined, the article author played a role in selecting the indexing in-

Table 3

Indexing Formats

Type of Indexing Information	Location of Indexing Information				Total of all Locations
	Title Page	Table of Contents	End of Article	Inside Cover	
Keywords	67	1	0	1	69
UDC Classes	54	0	0	0	54
PACS Classes	19	0	0	0	19
Index Words/Terms	8	0	0	0	8
CAS Registry Numbers	0	0	6	0	6
AMS Classes	4	0	0	1	5
CR Categories	3	0	0	0	3
CR General Terms	3	0	0	0	3
Subject Headings	2	0	0	0	2
Section Headings	1	0	0	0	1
Abbrev's/Trivial Names	1	0	0	0	1
OR/MS Classes	1	0	0	0	1
Total	163	1	6	2	172[a]

[a]There are 172 total items because four of the 162 unique journals accounted for in Table 2 each had two types of indexing and three of the 162 journals each had three types of indexing.

formation for that article. Others involved in supplying the indexing included editors, professional indexers, a translator, and the non-English language versions of translation journals. Sixty of 172 sets of indexing information came from the author only, and when an author worked alone, the task usually was to derive keywords. In the thirty-one cases in which the author's indexing selections were edited, the editor usually was the journal editor. When more than one person edited the author's work, it was some combination of the journal editor, copy editor, editorial assistant, editor-in-chief, and subeditor. A few survey respondents volunteered that editors variously suggested, deleted, modified, or supplemented the author's indexing contributions. For fifty-five of the 172 cases, indexing information (usually assigned from the UDC) was published in an English-language translation journal; this indexing information therefore came directly from the original-language text. There was an additional case of a translation journal whose editor reported that the translator assigned the PACS indexing. A final source of indexing information was indexers. Eleven of the 172 instances were in this category, where source indexing was supplied to journals either by Chemical Abstracts Service (for the assigning of CAS Registry Numbers) or by the American Institute of Physics (for the assigning of PACS symbols).

DO INDEXING SERVICES USE SOURCE INDEXING INFORMATION? Editors of indexing services, such as *Applied*

Table 4

Indexers and Vocabularies

Who Provides the Indexing?	Is the Indexing Information Assigned from a Controlled Vocabulary?			
	Yes	No	Unknown	Total
Article Author Only	11	39	10	60
Non-English Language Journal[a]	54	0	1	55
Author and Editor(s)	10	12	9	31
Indexer Only	10	0	1	11
Editor Only	2	1	1	4
Translator Only	1	0	0	1
Unknown	3	1	6	10
Total	91	53	28	172

[a]We found the source indexing in the English language translations of these journals.

Science & Technology Index and *Excerpta Medica*, reported that they and their indexers were more likely to use source indexing information as a guide to indexing rather than to disregard it or to accept it unedited. We received responses from twenty-three of the forty-four secondary services that each index at least one of the 162 journals identified as having source indexing. We asked

> . . . does the indexer use the source indexing information (keywords, classification numbers, subject descriptors) unedited? Does the indexer use this information as an indexing guide? Or is the source indexing information disregarded?

Most editors (nineteen of twenty-three) said that this indexing data served at least sometimes as a guide to indexing. That is, an indexer would not accept blindly a source indexing term but instead would consider this as one of several pieces of evidence about the content of a document. The reliance on this evidence varied among the indexing services. Seventeen of the nineteen editors implied that the guiding function of source indexing was not essential: "Our indexers use source indexing only to assure themselves that they have not overlooked any aspect of the article that is significantly discussed"; "source indexing information may be used as a guide, but in principle the editor selects appropriate terms independently"; "no indexing information is disregarded". Only two of nineteen editors indicated that the guidance provided by source indexing might be very important: "We look at the keywords, and adopt them as index entries if we think they are appropriate."

Four of the twenty-three responding editors adhered to a policy of not employing source indexing. Some comments were very emphatic: "We disregard entirely any source indexing information." Others were less so: "We do not use any of the source indexing information"; this case left open the possibility that the indexers could be guided by the source indexing information without "use", that is, adoption of the indexing.

DISCUSSION

USE OF SOURCE INDEXING BY READERS. The results of this study indicate that source indexing occurs frequently enough in some fields of science to permit readers to use this indexing information in a manner similar to the way they use the rather common title page abstracts. The abstract can serve two functions. First, the description of an article in an abstract helps a reader decide whether to read that article. Second, sometimes an abstract, especially an informative abstract, could supply enough detail of the article's findings to satisfy a reader and make unnecessary a reading of the article itself. Source indexing can serve these functions, too.

Because source indexing information can indicate the content of an article, the indexing has the potential to help both the reader with peripheral interests and the expert reader decide whether to read the article. For example, consider a reader of a journal like the *Proceedings of the American Mathematical Society*, which covers a wide range of topics in mathematics. By browsing through the pages of the September 1984 issue, one can look at the author supplied keywords and quickly see that the articles deal with local-colocal algebras, continued fractions, linear differential equations, and so on. It is true that looking at the titles of these articles would provide some of the same information, but there can be a difference between title words and keywords. The keywords can provide what some indexers call a more "exhaustive" analysis of the article than provided by the title alone; three, four, five, or more keywords could cover more topics, especially more minor topics, than does the title. From this issue of the *Proceedings* comes the article "Borel measurable images of Polish spaces". The accuracy of the title is verified by the first two keywords (actually, keyphrases) that are used as source indexing: "Polish spaces, Borel class of sets and functions". But the third keyword, "selectors", tells us that the article also involves a selection theorem.[21] Keywords might be most helpful to the reader who is not an expert in the primary topics covered by an article. A reader with peripheral interests might need more than a title written in the jargon of an unfamiliar subject; perhaps the reader would find his/her own specialty mentioned as a minor topic among the keywords.

Source indexing could even be informative enough to be a surrogate for an article. Consider a reader who is browsing through retrospective issues of *Applied Physics Letters*. The source indexing accompanying "Small-signal amplification in GaAs lasers" includes "GaAs laser diodes" and "77 °K". For some readers it will be sufficient to know that the authors operated their diodes at seventy-seven degrees Kelvin; reading the article will not be necessary.[22] For someone who already has read the article and wishes to recall this piece of information, the source indexing could save a rereading of the methodology.

INCIDENCE OF SOURCE INDEXING IN SCIENCE. In the sample of journals used here, source indexing appeared much more consistently in the physical and life sciences than in the earth and environmental sciences. An explanation for this might be that historically source indexing occurred first in fields like chemical engineering and physics, but that does not explain the high incidence of source indexing in the life sciences. Some might argue that the development of controlled vocabularies in the various fields would affect the use of source indexing. However, two of the major formats of source indexing, keywords and Universal Decimal Classification numbers, are applicable to many fields of knowledge. There are unanswered questions here, such as: was our sample biased? Are the purposes of journals in some fields different enough from those in other fields to affect the decision on whether to use source indexing?

SOURCE INDEXING FORMATS AND CONTROLLED VOCABULARIES. The use of both indexing assigned from controlled vocabularies like the Universal Decimal Classification and the Physics and Astronomy Classification Scheme and indexing derived as keywords suggests the difficult choice betwen controlled and uncontrolled vocabularies. One can expect that a controlled system is expensive to use: there are vocabulary lists to create, maintain, and publish; authors, editors, and indexers must take the time to search through the lists and assign their indexing terms. On the other hand, keyword indexing can be a very mechanical activity requiring one only to select indexing terms that are derived from the article itself. Therefore, a keyword indexing system might be attractive for author indexing; the author knows the article well and readily derives keywords. The use of both controlled and uncontrolled vocabularies might be a compromise, and, for example, some mathematics journals use source indexing terms taken from the AMS classification scheme as well as uncontrolled keywords selected from the articles.

AUTHORS, EDITORS, AND INDEXERS. The results of this study show that the author does play a major role in source indexing. The involvement of authors in the indexing of articles in more than half of the journals examined indicates that the author is a viable source of indexing information. Those who are concerned with the appropriateness of author indexing should note our finding that it is not unusual for authors to share their indexing responsibilities with one or more editors.

USE OF SOURCE INDEXING BY SECONDARY SERVICES. It is encouraging to find that indexing services do use source indexing information. And it is appropriate that this use is for guidance. Therefore, the indexers or editors working for indexing services can handle source indexing information without compromising quality control. An indexer still has the opportunity to analyze an article and provide indexing terms for it. If source indexing terms are available, the indexer can use the source indexing to: (1) check for concepts that the indexer might have missed; (2) obtain confirmation for concepts indexed by the indexer; and (3) obtain ideas for increasing the exhaustivity—or number of concepts covered—by the indexer. The source indexing even could hint at an important concept whose very importance and/or meaning is not expressed clearly in the text of the article. One difficulty for some indexers will be the translation needed between the language of source indexing and the language of the indexing service. For example, it will require skill to translate the common keyword of source indexing into the proper entry of an indexing service's controlled vocabulary.

REFERENCES

1. *International conference on science abstracting, final report*. Paris: Unesco; 1949: p. 59-60.

2. Morse, Rollin. Information retrieval. *Chemical Engineering Progress*. 57 (5): 55-58; 1961 May (p. 55).

3. Atherton, Pauline. *Aid-to-indexing forms: a progress report*. New York: American Institute of Physics; 1963: p. 1.

4. Morse; p. 56.

5. Speight, Frank Y.; Cottrell, Norman E. *The EJC engineering information program—1966-67*. New York: Engineers Joint Council; 1967: p. 6.

6. Herner, Saul; Johanningsmeier, W. F. Information storage/retrieval. Is it working? *Chemical Engineering Progress*. 61 (3): 23-29; 1965 March (p. 29).

7. Holm, B. E. Source indexing and vocabulary control. *Second National Symposium on Engineering Information*. New York: Engineers Joint Council; 1965: 16-24 (p. 17).

8. Speight and Cottrell; p. 5.

9. *Survey of the use of source abstracts and source index terms in a selected group of engineering journals*. Burlington, MA: Information Management; 1968: p. 2.

10. Price, Douglas S. Possible impact of electronic publishing on abstracting and indexing. *Journal of the American Society for Information Science*. 34 (4): 288; 1983 July.

11. Jennings, Margaret. The electronic manuscript project. *Bulletin of the American Society for Information Science*. 10 (3): 11-13; 1984 February.

12. Preschel, Barbara M. Q & A. *American Society of Indexers Newsletter*. no. 67: 10-11; 1984 May-June.

13. *Ulrich's international periodicals directory*. 23rd ed. New York: Bowker; 1984.
14. Sours, Keith J. *SCSS short guide*. New York: McGraw-Hill; 1982.
15. Foskett, A. C. *The universal decimal classification.* Hamden, CT: Linnet; 1973.
16. *Physics and astronomy classification scheme—1983*. New York: American Institute of Physics; 1983.
17. *Chemical Abstracts index guide 1984*. Columbus, OH: Chemical Abstracts Service; 1984: p. 89I.
18. 1980 mathematics subject classification. *Mathematical Reviews Annual Index*. 1983: S1-S34; 1983 December.
19. The full *Computing Reviews* classification scheme. *Computing Reviews*. 24 (1): 13-22; 1983 January.
20. Subject classification for the OR/MS index. *Operations Research*. 31 (6): following page 1210; 1983 November-December.
21. Levi, Sandro; Maitra, Ashok. Borel measurable images of Polish spaces. *Proceedings of the American Mathematical Society*. 92 (1): 98-102; 1984 September.
22. Crowe, J. W.; Craig, R.M. Small-signal amplification in GaAs lasers. *Applied Physics Letters*. 4 (3): 57-58; 1964 February 1.

SPECIAL PAPER

The Marketing Approach Applied to Special Libraries in Industry: A Review of the Literature

Maryde F. King

ABSTRACT. Consists of a review, with analysis and annotations, of the book, periodical and report literature dealing with the marketing of special libraries in industry, with particular emphasis on sci-tech libraries. The author also suggests related areas of research that would be useful if carried out.

1. INTRODUCTION

The first question that one may ask is why try to review and study the marketing approach as applied to special libraries in industry. There are many other varied special libraries and perhaps it would be just as useful to include all special libraries as to specialize and study only a segment of the group. One important reason for restricting this study to special libraries in industry is that such information is generally not available in this area and is needed. By using marketing terminology, it may be possible to improve communications with industrial management people concerning their own

Maryde F. King is Manager, Whitney Information Services, General Electric Company, P.O. Box 8, Schenectady, NY 12301. She has the B.S. degree (chemistry) from Whitman College and the MLS degree from the University of Washington.

special libraries. Special libraries in industry have sometimes been termed "orphans" since they do not fit easily into the organization charts. Another aspect of this orphan concept is that they may be included with the facilities services, where they are not understood, and they should be included with the intellectual services, where they are used.

It is frustrating and difficult for a Manager of Facilities to understand why a library has to pay several thousand dollars for *Chemical Abstracts* each year. He is much more accustomed to dealing with cases of paper goods and cleaning materials which he can shop around for and often get a good price break if he is careful and persistent. Library purchasing frustrates him exceedingly because if he makes arrangements with a vendor that gives a good discount for book titles that he needs to purchase, the service is generally so slow that the users and the library complain loudly about not receiving their materials. If he wants to stop the complaints, he usually finds that this can be done by paying greater prices for the books—and this destroys the advantage of the good discount arrangement.

Not all libraries in industry are in the orphan category since many of them report directly to the Director of Research, and usually this is a more favorable arrangement since the Director understands the relationship of the library to the researchers and their needs. Occasionally the libraries are located in the actual research department, and this is also generally favorable since the library and its users are under one management. Where there are many research departments, this becomes a more difficult situation as each department will want the library to purchase almost exclusively for them to the exclusion of others and it becomes a very diplomatic task to strike an equitable and fair balance in the purchase of diverse subject materials.

Academic libraries are not specifically included in this study since they derive both support and identification of their role through the accreditation surveys that are conducted for colleges and universities in response to their program offerings. This study also does not include the monetary or budget support of the special library as this is to be the subject of a state-of-the-art review by another author. This is essentially a philosophical survey of the marketing of the special library in industry.

To discover any references to marketing as applied to special libraries requires one to review marketing and marketing techniques. Since marketing is a concept, marketing terminology may not necessarily be used. This requires then that the basic concepts of marketing be kept in mind as the articles and reports on special libraries are studied. Up-to-date marketing is very complex. The main elements of marketing include product quality, packaging or styling character, the right price for the particular market, and advertising that communicates clearly and motivates consumers. In addition to these elements, distribution and display of materials are important. Distribution and display are extensively used in a library to make its products accessible. The main elements of marketing frequently utilized by a library are product quality, the right price, and to some extent, advertising. Packaging or styling character are often lost in library growth and space problems. Library products rarely directly reach budget decision makers as library information is often integrated by managers into their reports to these people so that they never know the actual product or its quality.

The literature surveyed for this article included the government report literature, journal articles, proceedings of seminars, conferences and symposia and the monograph literature dealing with special libraries in the industrial setting. A rather broad area was studied in order to see if it was possible to locate direct marketing references in relation to special libraries in industry. If there were no direct references to marketing, then other terminology was used because the subject might be completely disguised or integrated in a study reporting some other aspect of the industrial special library.

2. DEFINITIONS

Marketing terminology is rarely used to sell library services. The function is disguised under a number of other descriptions since users are not willing or able to place a dollar value on information. To paraphrase a marketing definition to fit library services, "Library service for the user is the right information for the right person at the right time at the right price for the users' needs." To the user seeking information, the price is the time to go to the library and phrase the question to the librarian—or maybe just to use a book from the reference shelf. The question to ask ourselves is: What is the price if there is no library or only an inadequate library? In the most expensive case, it may be the price of the time and materials it requires to do an experiment to obtain the information. If this is the only manner by which the information may be obtained, it may be essential to get the information in this way. However, if the information is fairly readily available in a standard library reference volume, the "price" is definitely too high for the information obtained by experiment. If the amount of time required to locate the information in the literature is too great—then once again the "price" is too great. Too much time searching may also defeat the marketing criteria of having the right information at the right time for the user. There may be another variable in the overall picture. In the research situation, it may be impossible to judge the timing, the information, and the person. Seemingly useless information may suddenly acquire great value during the progress of research. Laboratory curiosities become effective links in development work. This aspect of information has been described as follows by Martyn Clemans:

> Information is a curious commodity. Although it can be bought, sold and exchanged, only its outward manifestations can be perceived by the senses. The thing itself is a will-o-the-wisp, given life only by what the mind already knows.[1]

Users of sci-tech special libraries are decision makers whether they are scientists or managers. They require information to do their work and this information may be obtained through library services.

The "product" is the area of service for the industrial library user. The areas of service for industrial libraries may be grouped as follows:

1. Libraries serving research and development personnel.
2. Libraries serving production facilities, including limited research on production problems.
3. Libraries for management information services.
4. Combinations of the above libraries.

Marketing this "product" of library service is every library staff member's job but is usually not included as a statement in the job description.

3. REVIEW OF ARTICLES ON MARKETING SPECIAL LIBRARIES

Studies concerning the "marketing" of special libraries are usually disguised as special studies on the need for a library facility or library services. The opening sentence in Saul Herner's and M.K. Heatwole's[2] article on research libraries reads, "The need for scientific and technical research as an adjunct to military activity has long been recognized and implemented.'' They then indicate that there may be subtle differences between military and nonmilitary research libraries. The primary difference is considered to be expedience. More extensive and more rapid service is required by military library users. Libraries must be organized to give rapid access to their materials, and the staff should be technically trained to understand the user's problem and provide correct and useful information. Another difference in the military research library is the task or mission orientation. This usually dictates a ''tight'' collection that concentrates on recent materials that are directly related to the subject of the mission. A third differentiating feature may be the necessity for security for classified documents. Physical layout and costs are then discussed for the best and most economical plan for a military research library. This is truly a marketing article emphasizing the current materials as the right information, organized for expedient availability (the right time), and economically housed and maintained (the right price) for the user.

Needs and habits of scientific authors and readers are discussed by the Blaxters[3] in an article in *Nature.* Their discussion opens as follows: ''There is little doubt that the cost of providing scientific information is increasing.'' Costs of scientific and technical journals, for example, have increased 48% in six years, while medical journal costs increased 64% in the same period of time. Science libraries studied by the Blaxters indicated that costs for materials purchased for the libraries more than doubled in the six-year period. The effort here was to define the operational needs of the scientists in an attempt to learn the optimal size of the library. Once again the needs of the user and the size and cost of the collection are the marketing criteria, or the right product at the right price.

By emphasizing the right product at the right time, we find P.E. Colinese[4] saying that to keep in touch successfully, an information department must be dynamic and so organized that it is sensitive to the user's changing needs. Information needs are of two types—service needs, such as abstracting, indexing or cataloging and other library services, and subject needs supplied through reference or bibliographic services. Provision should be made for attendance at research planning meetings by some member of the library staff. The staff should be aware of the organization's R&D program—not just through the research reports, for they deal with work already done, but through information concerning on-going projects. In addition to this internal information, one must also watch for new branches of science as they develop outside the organization. Changing user needs will require changes in services, especially if the library resources of space, money, and manpower are inelastic.

Colinese includes, as a user's need, the necessity to receive adequate publicity about the services offered by the information department, since it is possible for potential users, particularly in large organizations, to be ignorant of the department's existence.

Thomas J. Allen and Stephen I. Cohen[5] in their study on "Information Flow in Research and Development Laboratories," have examined the individual engineer and his performance with respect to the phenomenon called information flow. Starting with the hypothesis that no laboratory can be completely self-sustaining and must import information from outside, they indicate two ways of doing this. The literature can be used by the staff to keep informed about recent developments in the field, or knowledgeable people can be consulted for this purpose. Two research and development organizations were studied to determine the information flow patterns. Two classes of individuals seemed to be evident. One group had few contacts outside the organization. This group was the larger one. A much smaller group had extensive outside contacts and served as sources of information for their colleagues. This latter group has been described as technological gatekeepers. Technological gatekeepers vary in the type of information sources used. Some rely more upon the literature, while others rely on oral sources. There are two important implications for management from this study: first, the flow of technical information should be understood in order to improve the communication system; and second, the value of gatekeepers should be recognized. Gatekeepers should be allowed easy access to the literature. To minimize cost in providing access to search and retrieval systems, access should be given first (and perhaps only) to the technological gatekeepers. The need for the product at the right price is still the primary concern for the user, who in this case is a communication link to other users and is described as a gatekeeper.

The need for an industrial library is described as follows by E.B. Uvarov[6] in "Starting a Small Industrial Library." Uvarov begins by stating that scientific books and journals are used in all laboratories. These items form the nucleus of a library in an informal sense, even if there are only a few books purchased by the laboratory. These books may be held by individuals, and two or three journals received as a result of personal memberships in learned societies form the library nucleus. The industrial library starts when the need arises for a more abundant and less haphazard supply of literature, and when that need becomes an active and urgent demand, the time comes to provide a recognized library. The various steps are described for starting the library including the place, the product (library services needed), the price (library budget), and promotion (policies and growth). Good marketing requires planning, and in the conclusion we find the following statement:

> Intelligent planning, a sensible policy and a good library will ensure that the library is a valuable asset and not an expensive collection of waste paper.

A 1973 United Nations publication entitled "Information Units in Small Plants"[7] contains two major developments concerning information service. The first of these is greater emphasis being placed on anticipating the user's needs and, second, more appreciation of the value of speed, accuracy, depth and economy in finding the information required. Three elements of marketing are prominent in these developments—product, price, and promotion and, even though not mentioned, place must be understood. These developments require close interaction between supplier and user of information, and demonstrate the need for a higher calibre of library staff for this type of work.

By contrast, the report describes reference service as a means to hoard material and to investigate specific requests only. "Communication has been called the life blood of industry" according to this report. The UN report continues by saying that the life blood is too often in short supply. The information service is described as it evolves in a small plant—when the business starts there are only a few people and a memorandum, word of mouth, or sharing a relevant article with a co-worker may take care of necessary communications. Business growth and increased specialization make it difficult to continue the informal methods of communication. Individuals have less time to indicate interesting items to colleagues and, in fact, in scanning the literature they become blind to items that do not directly concern their business activities. This is described as the critical stage when much valuable information is passed over and when a firm should realize it is time to install an information unit. It is then essential to design an information unit. The elements of marketing are all here—the right product (information service), in the right place (the developing business), at the right time (to avoid communication breakdown or lack of communication), and at the right price (the information unit designed to meet the needs or requirements of the firm). If the information unit is given the right promotion to achieve its objectives for the firm, it should have:

a. An intimate knowledge of the interests and current operations of the firm and of each member of the technical staff;
b. A library of its own and access to other libraries;
c. Contact with other sources of information; e.g., firms, institutions and individuals;
d. The ability to exploit these resources effectively.

Complications that may arise in attempting to do a proper marketing job with technical information have been studied by Dr. George Anderla in "A Challenge for Governments and Society,"[8] published in the April 1973 issue of the OECD Observer. This study deals with the technical information field over the next fifteen or twenty years and forecasts the skills and numbers of information specialists that will be required. The study was undertaken because information is now recognized as a vehicle for the transfer of knowledge and a basic resource. It is considered an essential ingredient for the decision making process and production processes of all kinds. Dr. Anderla makes three points: first, the generation, transfer, and use of information problems have been largely underestimated as far as real dimensions, complexity, and growth dynamics are concerned; second, information is inadequately handled and managed by a multitude of mutually competitive fragmentary bodies using outdated tools; and third, the extraordinary dynamics of information require vigorous government action using appropriate technologies to make use of this valuable resource for the development of our future society. He then develops his forecast concerning the growth and automation of information. His ten conclusions cover the implications of the growth and automation of information for industry and government as we approach the turn of the century. In one of his conclusions he says, "It also means that an 'information market' as wide and open as possible must be created and institutionalized or in other words that interface between information supply and demand must be rationally organized." Dr. Anderla closes by saying that not all will agree now with his forecasts or conclusions; however, he feels that that is only a matter of time.

The Educational Resources Information Center/Clearinghouse on Library and Information Sciences published "The Marketing of Information Analysis Center Products and Services in June 1971."[9] In the summary and conclusions, it was stated that information analysis centers represent a valuable national resource which has not been fully utilized. Government economy measures require reduced federal support for these centers; therefore, service charges have been introduced. To evaluate the marketability of the information service, the center must consider the following: national policy, center users, specific publication or service, channels of distribution, advertising and sales promotion, and pricing.

The authors concluded that the information centers must educate users and managers on the cost of providing information and the value obtained by using the Information Analysis Centers, with the final decision on service charges being in the national interest.

The marketing of special libraries is of concern in industry, in national governments, and on an international basis. On the 10th of June 1970, the Royal Society held a one-day conference on "International Developments in Scientific Information Services." In Dr. F.A. Sviridov's paper, "International Cooperation Among Information Specialists: The Work of FID,"[10] information services are considered mandatory. Dr. Sviridov says, "Rapid progress of science and technology and its growing influence on all aspects of life of the modern society are characteristic features of our time. Well-functioning and effective information services are prerequisite for any further development (from preparing a small experiment in a research laboratory to decision making by a government agency." In this case, the product is assumed sold and the article then goes on to detail the right information, in the right place at the right time and price for the user.

Scheffler and March study a government information center in "User Appraisal and Cost Analysis of the Aerospace Materials Information Center."[11] In evaluating this system, marketing criteria were once again employed.

The information system was evaluated regarding its performance of fulfilling its primary mission of serving the information needs of the users (the right product) and of improving its efficiency in serving these needs (right time and price). The services provided to the user should correspond as nearly as possible to user requirements. The following points were brought out in the summary: one, that information services are expensive to maintain and two, that although some cost figures are available, it is very difficult to establish the corresponding cost savings effected by providing timely, pertinent information. Information services are usually considered expensive overhead without taking into account their overall economic and other advantages.

James O. Vann, in describing the Defense Documentation Center (DDC) for Scientific and Technical Information[12] some years ago, emphasized five points of importance that were implemented by the instruction describing the DDC mission. These five points speak for themselves as they are described below.

1. The acquisition should be made of *ALL* Department of Defense scientific and technical documents by DDC, which will make it *The* Defense Documentation Center in the true sense of the word.
2. Prompt and well-indexed announcements of newly acquired DoD scientific and technical information documents should become a specific requirement. Manpower and an advanced electronic data processing system are provided to support this effort and will aid in decision making in the selection of documents by engineers and scientists.
3. Timely dissemination of scientific and technical documents to the DoD community is one of the specific objectives. Authorization of this function will speed DDC secondary distribution and ultimately improve primary distribution.
4. In addition to the documents and their distribution, a clearing house is to be maintained in the form of an index of current research, development, test, and evaluation programs.
5. DDC is also to establish a centralized directory and provide referral service on available DoD-sponsored scientific and technical information which will coordinate DoD Specialized Technical Information Centers as part of an integrated DoD system.

The new DDC instruction of 1963 presents a challenge to carry out these responsibilities with minimum resources and an opportunity to provide very tangible assistance to the nation's defense engineers and scientists in their race to keep our overall lead in science and technology.

The points emphasized by Vann are all directed to improving services through better and more prompt access to the United States government document literature.

In 1967, the North American Aviation Company analyzed the technical information requirements of the Federal Aviation Agency in bulletin AD 651926.[13] This document, entitled "Analysis of Scientific and Technical Information Requirements of FAA Contract No. FA64WA5213," anticipates significant changes in information handling will be evolving over the next several years. These changes are likely to include more centralized data handling as information storage, retrieval, and communication methods become more sophisticated and relatively less costly. By 1970, the optimum concept for FAA may be a centralized technical information system providing an awareness service, reference and research services, interlibrary loan service, and files on each FAA library user and FAA library system document. Increased interaction is expected with other government agencies, such as NASA and DDC.

This study then predicts over a longer time span—10 to 50 years—the nature of libraries is expected to change to an emphasis on information rather than on documents. It goes on to say that libraries will be expected to provide briefs of specific subjects rather than a stack of documents about the subject. This service will involve considerable expenditures for automatic data-handling systems. These expenditures will probably limit such installations either to one central facility or to facilities where the technical information handling function shares equipment with other data handling functions.

The report further emphasizes that technology including the field of aviation is becoming increasingly complex. New fields of specialization are occurring almost daily. An individual reader, to keep abreast of developments, must be continually digesting new scientific and technical information. For the FAA, it was stated emphatically *that if the FAA wishes to maintain its preeminence in the field of aviation, it must take an aggressive position with respect to seeing that its personnel keep up-to-date on technical developments.* An unconditional prerequisite to this is an active technical information program internal to the Agency. It was also estimated on the basis of job assignment that 85% of all Agency personnel were potential library users who need to use a library to keep abreast of technical developments.

There were many reasons for non-use of the library including distance. However, the data showed unmistakably that when a library was established, personnel in the Agency made use of it. This was a very strong marketing report, emphasizing the user's needs.

The Committee on Scientific and Technical Information (COSATI) of the Federal Council for Science and Technology has titled its annual reports as follows: "Progress in Scientific and Technical Communications."[14] COSATI has continual concern for the evolving scientific and technical systems in this country; the ability of Federal agencies to utilize scientific and technical information effectively and efficiently in carrying out their missions; development of internationally compatible information systems leading to maximum interchange of information and development; and application of information processing technology, including telecommunication, satellites, and computer science. These continuing concerns are all marketing concerns—emphasizing only a portion of the total marketing picture. COSATI's concern is mainly with packaging, availability, and utilization. Other concerns of COSATI are government agency information and communication systems, cooperation and interrelations between agencies, as well as cooperation between agencies and private organizations, and international efforts.

An important study during 1969 known as the *SATCOM* report, dealt with scientific and technical communication of the National Academy of Sciences.[15] It examined the nature of the problems facing the information generating and using communities in both the private and public sectors. The costs to both the generating and using communities were a major concern, as well as the quantity of material to be handled.

In *The Knowledge Revolution*[16] by D.N. Chorafas, there is a chapter entitled "The Knowledge Industry." Here Chorafas relates the following story about Ptolemy's library at Alexandria. This library was conceived on a tremendous scale. Whenever anyone brought an unknown book to Egypt, he had to have it copied for the library. For dissemination purposes, a considerable staff of copyists was engaged continually in making duplicates of all the more popular and necessary works. This resulted in Alexandria's library being in the bookselling business. Callimachus, head of the library, arranged for the systematic arrangement and cataloging of this voluminous accumulation. The library attracted scholars from all over the known world. Ptolemy II ingeniously offered these learned men twice the salary earned in their home countries if they would stay and work in Alexandria. Here users flocked to the library services in somewhat the same way buyers flock to a good sale. It has been said that a good restaurant does not have to advertise, and yet it will always be busy. Ptolemy's library in Alexandria was in such a position. The size and scope of the collection with Callimachus' organization meant that the right information could be found for the user. The place and the price were right for Ptolemy who took advantage of the attraction of his library for scholars to employ them for the benefit of Alexandria.

ENDS—European Nuclear Documentation Service,[17] described by Carl O. Vernimb and Gunter Steven, states that an important reason for the trend toward automation in information is the fact that our technological society has a greater need for information today than it ever had before, and that the quality requirements imposed on a documentation service in terms of speed, exhaustiveness, and precision are increasing every day. ENDS was developed by the Centre for Information and Documentation of the Commission of the European Communities and has been in operation since 1967. This service is to be kept at the highest possible technological level and is to serve as a basis for research and development in the field of information technology. It is both an operational system for the benefit of users and an experimental system at the same time. Two important elements of marketing were of primary concern in developing the system for the users, and these are the right material within the correct time.

James Hillier of RCA Laboratories wrote in "Measuring the Value of Information Services"[18] that he found a study of the literature on the subject both confusing and disappointing. The problem was vaguely described as "satisfying the user's need for information" without defining the need or appearing to recognize the extreme ranges of variation that can exist in both the nature of the user and the nature of the information. Mr. Hillier continues by saying that the first task, from the point of view of management, is to establish the nature of the need. A laboratory engaged in true exploratory research can be considered the "worst case" for establishing the nature of the need. He then gives a list of five major needs in the order of increasing importance to the laboratory. Costs are not considered here. The information flow must:

1. Prevent excessive duplication of research. This can be best achieved by the information service that is an integral part of the technical team.

2. Provide specific information needed by the technical staff.
3. Provide "catching-up" information for the individual user who finds it necessary to become familiar with a new field.
4. Provide an efficient means for enabling members of the technical staff to "keep current" in their fields.
5. Stimulate creative thought in a way which will maximize the probability of occurrence of creative ideas that are valuable to the company. Since the generation of creative ideas is the basic reason for the existence of any laboratory doing research of any level or kind, and since information is vital to creative thinking, this is the most important requirement of all, according to Mr. Hillier.

In conclusion, Mr. Hillier said that it was clear that we do not have enough detailed and quantitative data to enable management to make a reliable evaluation of an information service activity in a specific laboratory, particularly when that laboratory is a heterogeneous one doing exploratory research.

The user's needs were well described, and timeliness and availability were inferred. Since the needs were essential, costs were not considered.

C. Allen Merritt and Paul J. Nelson in the ITIRC-003 report "The Engineer-Scientist and an Information Retrieval System"[19] give a very straightforward marketing statement. ITIRC, the IBM Technical Information Retrieval Center, was established to serve the IBM scientists and engineers in all their laboratory locations. ITIRC exists to supply the right information to the right person, in the shortest possible time and at the least possible cost. Merritt and Nelson go on to say that this is no small undertaking, when one considers the size of the company, the number of locations, and the tremendous range of interests in research, development, manufacturing and sales. Users range from physicists to circuit designers to programmers. The need for an information retrieval center may become urgent. However, it cannot be established until management recognizes its value and is willing to set aside funds and manpower for its operation. The report indicates, further, that the funding must cover a period long enough to permit a valid judgment about the value of the services rendered. The report then covers the dissemination, announcement, searching, and microfilm information retrieval systems at IBM. These are evaluated on the basis of the marketing statement above. This retrieval system is backed by the technical library. This "marketing" report covers only one part of the total library service. The marketing statement, however, covers the spirit of the total technical library service.

In November 1972 William T. Knox delivered a speech entitled "Technology Transfer Failures and Successes." A later article in *Science* of August 3, 1973, was adapted from this speech.[20] This is not a marketing article but rather an article on the need to study the problem from a marketing point of view. In spite of spending several billion dollars to develop and operate better technological information systems, users are still dissatisfied with the effectiveness and the responsiveness of existing systems. Technological information transfer is the dissemination of what Fritz Machlup called practical knowledge—knowledge useful in one's work, decisions and actions. By looking at this problem with marketing strategy in mind, the approach of having the right information, in the right place at the right time, and at the right price—all this could provide more effective transfer of technological information.

The recent book published by Greenwood Press on *Public Relations for Libraries*, by Allen Angoff, includes a chapter on "Public Relations in Special Libraries" by Elizabeth Ferguson.[21] In a well stated brief statement, she says, "Special libraries exist to expedite the process of locating and providing, from the present mass of available print, specific knowledge that is needed on the job." Another way of expressing the market emphasis is stated in her purpose for the special library which is to save the time of expert personnel, save company money in buying printed matter, and to produce better information more efficiently. She goes on to describe public relations for the special library as composed of two elements; one is to do a good job and the other part is to tell about the service. This is further described by stating that the library provides the materials and reference service for essential research in the conduct of the business which the special library serves, and that the library promotes its services so that it is looked to for assistance. Although that article deals mostly with public relations, it also includes some excellent "marketing advice" on how to achieve good public relations.

Alan Armstrong recently wrote in the *Financial Times* for December 17, 1973, that profitability depends on getting accurate, relevant information to the policy makers.[22] He cited that the executive's biggest timewaster is inadequate or inaccessible information. In his article entitled "For Good Decisions You Need the Facts," he noted that the remedy was a business library. Mr. Armstrong observed, "Just as industry set up libraries in the 1960's to harness the technical information explosion, it is creating business libraries today to cope with the flow of commercial information." Both the marketing and monetary values of the library are well described in the following quotation:

> The link between the executive's desire to shed the information burden and the business librarian's ability to handle it, is financial. Costs can be cut by preliminary literature searches to prevent costly market research which would otherwise duplicate published findings.

Without using marketing terminology, Mr. Armstrong states the case for marketing the special library very effectively simply by saying that up-to-date information is a vital component of every business operation.

Almost twenty years ago, J.C.R. Licklider tried to forecast the direction that libraries might take in "Libraries of the Future."[23] Much research and study went into that report and it is not a marketing report. However, it does contain one key statement which can be rephrased as a marketing statement, namely that information be available when and where needed. Much of the book then deals with how this can be achieved. Licklider's work deals with the product that we want to market—how it should be packaged, what is the most useful format, and how it should be handled and accessed. Marketing itself is a big and responsible job and we need the right information at the right time—but we also need it in a marketable format, and that is the emphasis that is found in Licklider's study.

Up to this point, the marketing of the special library has dealt with the delivery of library services to the user. In the AGARD Lecture Series No. 44 on "Scientific and Technical Information Why? Which? Where and How?",[24] H.A. Stolk in his paper on "Sources of Scientific and Technological Information" includes a section on the user and his behavior. Stolk finds three major audiences who use technical information: the general audience, the mission audience and the technical management audience.

Technical information used by the general audience is characterized by the fact that the generator of the information does not know specifically who will use the information or when. Use of this information by the general audiences may be for purposes totally different from those of the generator. The mission audience is characterized by a close coupling to the generator of information. This close coupling between the generator and user results in efficient information transfer. However, we also find that the information tends to stay within the relatively narrow confines of the generator-user environment although it could be of use to the general audience or other mission audiences. The technical management audience needs timely and accurate technical management information systems because of the increasing expenditures on research and development along with the additional complexity of the programs. Systems developed for this audience are also employed by working engineers and scientists who use the system to identify on-going research and technology efforts related to their particular areas of interest. The technical information content in this system is minimal but is sufficient to determine whether the performer should be contacted for detailed information. One problem encountered by users is that they need engineering type information (performance characteristics, test data, etc.), items which the formal library systems can hardly provide because they have not been organized for this purpose. There is also some indication that scientists and engineers have not learned to use libraries and information systems in the most efficient way. The Parry Report states from a sample taken from twenty-three British Universities that: *only 37%* of undergraduates know what abstract services are; *only 14%* have been taught to use them; *25% do not know* that their library has an author or subject catalog; and 41% do not know there is an interlibrary loan service. This seems to indicate that users need some kind of training in the use of libraries and information services.

Stolk goes on to say that another important feature is: most users like to have their information needs met instantaneously. What frequently happens is that the user makes a quick minimum effort at getting information. If the optimum information is not found during this first try, he will too often resort to the use of readily available but less than optimum information. For example, an engineer selecting materials may not use a low cost material because he cannot readily determine its characteristics in a particular environment. Instead he picks an expensive alloy which he knows will do the job. This gives rise to one of the frequently used arguments against expending resources to provide better technical information systems—the users seem to do their job without them! However, the real question is: "How could their performance be improved by instituting better technical information systems?"

A number of years ago, H.P. Luhn wrote an article describing a

Business Intelligence System.[25] That article dealt mainly with the development of an automatic system to disseminate information to the various sections of any industrial, scientific, or government organization. Data processing machines were to be used for auto-abstracting and auto-encoding of documents and for creating interest profiles for each of the "action points" in an organization. Both incoming and internally generated information is automatically abstracted and sent to appropriate action points. The system identifies known information, finds who needs to know it, and disseminates it either in abstract form or as a complete document. In "selling" this system, the following thesis was presented:

> Efficient communication is a key to progress in all fields of human endeavor. It has become evident in recent years that present communication methods are totally inadequate for future requirements. Information is now being generated and utilized at an ever-increasing rate because of the accelerated pace and scope of human activities and the steady rise in the average level of education. At the same time the growth of organizations and increased specialization and divisionalization have created new barriers to the flow of information. There is also a growing need for more prompt decisions at levels of responsibility far below those customary in the past. Undoubtedly the most formidable communications problem is the sheer bulk of information that has to be dealt with. In view of the present growth trends, automation appears to offer the most efficient methods for retrieval and dissemination of information.

The objective of the system is to supply suitable information through effective service at the greatest convenience for the user. Here we find a strong marketing statement for information service.

The case for management information systems has been expressed as follows by R.N. Kashyap in "Management Information Systems for Corporate Planning and Control".[26]

> Management information systems play a vitally important role in the development of corporate long-term as well as short-term plans and in achieving effective control of business activities at all levels within the enterprise.

Information is an important resource and may be considered equal in importance to the traditional resources of men, money, materials, and machines. Kashyap states that the availability of pertinent, accurate, and timely information is most essential for making effective planning and control decisions. The quality of decisions made by management is dependent on both the experience and judgment of the decision maker and the scope and accuracy of information available to him for decision making. Also contributing to the need for useful information is the increasing size of organizations and the mounting complexity of doing business. Risks of losses from poor decisions and opportunities of gain from good decisions have become very large. This makes it almost mandatory that the best possible information be made available for making the planning and control decisions in a business enterprise. Management information can be a dynamic tool if marketed as an efficient, accurate, and timely product for the user.

Henry Mintzberg in his study of MIS claims that the manager is in conflict with most formal information systems. In his "Myths of MIS",[27] he says that while the manager seeks trigger, speculative, current information the formal system usually gives him aggregated, precise, historical information. Also, MIS deals with internal information and the manager often needs external information. The first element in an effective MIS is that the manager must be provided with relevant information in a way that economizes his time. No doubt, much of the information provided in such a system will concern external events and will emphasize the current, the specific, and the uncertain. MIS also requires a more systematic and effective information disseminating system. The key characteristic of the disseminating system is speed. Mintzberg concludes that the manager, not the computer, is now the real data bank of organizational information. The information he needs and uses is not written and therefore not available to the MIS specialist. The user's need for information at the right time and price are still the governing criteria for information as well as formal information.

The development of a management information system is a long and costly undertaking, according to Larry D. Stout in his article on "The Problems of Management Information Systems: Why They Fail."[28] He summarizes the reasons for their failure as simply due to the fact that the systems do not provide the proper information, at the proper time, to the proper individuals. Mr. Stout believes that a statement of goals and resources would permit better planning. It is also necessary that users, management, and systems personnel should communicate and participate in the planning for the successful evolutionary development of a system. This effort is required so as to reverse the trend of failure and to produce the desired results of information and timeliness.

John E. Gessford, in his articles on "Management Information Systems Development,"[29,30] has said that for an MIS improvement to successfully take root it must be wanted and appreciated by the management and staff personnel who will use it. Thus the best development step also depends on the perceived information needs of managers. He goes on to say that information systems and models can sharpen the definition of corporate goals and policies and clarify their feasibility and implications for corporate stability and growth. Aside from the direct value of clarification to management, MIS provides a better framework for developing program planning information systems.

An example of some services that can be provided by MIS are statistics, based on historical marketing data; means of logically integrating marketing data; and the judgment of marketing experts and estimates of the value of investing in specific market research projects to get better information for sales response estimating.

The need was emphasized for a better marketing system through better information acceptable to management.

According to Professor Hax,[31] many managers see a need for more effective, mechanized information systems within their firms. The main objective of a management information system is to provide managers at all levels with supporting information to help plan operations and measure actual results against those plans. In planning this MIS, a modular, evolutionary design schedule then is proposed for the basic information systems. An evaluation of the cost and benefits of an improved MIS was included. An attempt was made to quantify specific benefits so that the impact of the management information system can be evaluated. It could be shown in this case that actual cost savings resulted from the improved operational and control data provided by this management information system.

Nicholas Long, in his article on "Information and Referral Services: A Short History and Some Recommendations,"[32] takes an interdisciplinary look at information. A new kind of social service has become known by the name "information and referral." Its purposes are to facilitate client access to human services and to obtain data for the planning of human services. This is another application for the management information system. The information and referral centers possess data that could be used to stimulate coordination; however, their primary function is facilitating access to services. This information service is not evaluated, but the need has existed in the social service area since 1870.

"An Evolutionary Approach to Marketing Information Systems"[33] postulates that the ability of an information system to respond to the evolving needs of its users is a necessary ingredient for its success. INF*ACT is an evolving and heavily used marketing information system. INF*ACT (Information for Action) grew out of the need for a system which would supply a capability for easy retrieval and analysis of data. This system is an on-line, time-sharing system used to retrieve and manipulate data in very general ways as dictated by highly individualized and changing user needs. Users who are not computer technicians may easily utilize the capabilities of the computer with minimal learning time. Routines in this system can easily be updated to meet evolving needs. Evolutionary capability was and continues to be the key concept for INF*ACT. It is a user-oriented system that works effectively and is a strategically correct way to attack the marketing information problem.

Bank management needs information just as various other managements do. A bank or other organization has its chosen objectives: the direction to which its management philosophy and performance requirements, its organizational structure, and decision-making processes are geared. Timely information necessary for self-preservation of the organization requires that it be aware, not only of where it wants to go, but also of where it is. The organization should be alerted as soon as possible to changes in environmental or internal conditions. According to Leslie Sloane in "Management Information I.",[34] a manager may be defined as a user of resources in the light of information. Since the provision of information absorbs considerable resources, a bank must be reasonably assured that it is going to improve the bank's profitability more than in proportion to this added cost. A management information system defined by Sloane is an organized method of providing past, present and projected information relating to internal operations and external intelligence. It supports the organization by furnishing uniform information in the proper time frame, to assist the decision maker. Information does not replace management; it is its essential tool. If management information does not improve future performance, it is not management information. Sloane repeats the familiar statement that management information should be the right information, presented to the right level of management at the right time, and in the right form. Concerning the cost of data, Sloane reports that the cumulative cost of providing data increases with volume, the degree of urgency and the standard of accuracy required. The ideal is neither to minimize costs nor to maximize information, but to optimize information value in terms of relevance of content, timeliness, and degree of accuracy required.

William R. King and David I. Cleland describe SPDIS in "Decision and Information Systems for Strategic Planning."[35] The authors maintain that effective strategic planning can be carried on only through the development and implementation of a strategic planning decision and information system (SPDIS). This concept includes the Management Information System as part of an overall management and organization plan. The authors believe that SPDIS is truly a "management" information system in that it focuses on the support of the strategic decision-making aspect of the manager's job and emphasizes the critical role of the manager in developing and effectively implementing strategic organizational planning.

The question of location of an information systems department is considered by W.E. Reif and R.M. Monczka in "A High Level Independent Location for the Information Systems Department."[36] A growing number of writers and practitioners have become convinced that organizational considerations are as important, if not more important, than technological considerations in the successful development and implementation of computer-based information systems. A study of the location of the information systems department in ten organizations confirmed the authors' belief that the high-level location will provide the greatest return on the company's investment in computer-based information systems.

Reif and Monczka give the following dominant factors favoring the independent, high-level location:

1. A company-wide view of information systems needs is required, as is the establishment of objective criteria for systems development.
2. Organizations need to take an integrated approach to systems design, implementation, and evaluation.
3. Greater top management involvement in and support of infor-

mation systems development is required to assure optimum results.
4. Increased acceptance, cooperation, and support from user departments exist when information systems departments are independent and high-level.
5. Effective utilization of information systems personnel, including the forming of task teams with the necessary skills to identify new system application areas and solve information systems problems, is furthered by independent, high-level information systems departments.

The high-level location is not only more effective, but should give the greatest return for the company's investment in this location.

In addition to location, there is the problem of communication. H.A. Burgstaller and J.D. Forsyth in "The Key-Result Approach to Designing Management Information Systems"[37] discuss the organization and communication of information. Specifically, they ask the question, "What information should be produced by management information systems?" A further question then developed by the authors was, "With respect to the management cycle of planning, executing and reviewing in those areas in which performance is critical, what classes of information do you need?" Information that relates to this question is management information. It becomes a resource needed in the same sense that money, manpower, and machines are required by a manager. In other words, information is required in order to get results. This management information system is described as the Key-Result-Areas approach. This approach concentrates on critical areas of performance and is concerned with selecting a relatively small number of information needs. This reduces the complexity of the resulting information system design for satisfying these needs. The information system remains effective in terms of the functions it is designed to serve by virtue of the fact that it relates to critical areas of performance. Managers involved in identifying information needs are more receptive to having information systems designed and implemented for their use. The authors' conclusion is that the Key-Result-Areas approach works.

Management Information Systems are a special area of library service to industry. The same marketing criterion holds for the right information, in the right place at the right time and price for the user in MIS systems. The idea is expressed in many ways, often not mentioning a single marketing criterion only emphasizing the need and how to satisfy it.

In 1963 the Organization for Economic Cooperation and Development held a Conference on The Communication of Scientific and Technical Knowledge to Industry. The proceedings were published in a book of the same title.[38]

The conclusions from this meeting were summarized by Professor B. Rexed as follows:

> If science and technology are to spur economic growth, they must be applied to industry; and if they are to be applied, they must first be communicated to and within industry. Information must be considered as an indispensable factor in science policy and it is essential that governmental and inter-governmental authorities assume their share of the responsibility for ensuring that effective information facilities are available to industry.

Professor Rexed goes on to say that restricted or unpublished government reports on research findings should be made available to industry. He suggests that industry could be instrumental in improving the quality of publications by being more selective in its acquisitions and its choice of advertising media. An additional recommendation was the establishment of a referral center in each country which would be beneficial both nationally and internationally by assisting engineers to locate the most competent sources of information and reference. The market for information exists, the product, the place and the price were the elements of discussion.

Gerald Jahoda wrote a thoughtful article on "Special Libraries and Information Centers in Industry in the United States."[39] He describes the changes in information services for industry in the U.S. during the early 1960's. Jahoda says that the reasons for supporting information services are similar to the reasons for support of research a number of years ago. Information services are considered a worthwhile investment even though the value of this activity cannot as yet be measured concretely. The increased work-load of the special library caused by requests for additional services, by greater use of existing services, and the increased size of the literature, have forced the individual special library to reappraise its functions, services and operations. For maximum service at a reasonable cost, the special library is making greater use of outside information services, increasing interlibrary loan cooperation, using data processing equipment for clerical operations and appointing specialists in information work. Library services are in demand and the costs are kept manageable through cooperative services and using data processing equipment.

The Nelsons in "Library Systems and Networks"[40] declare that the market for library services is nonprofit. Library service has the disadvantage of not having the yardsticks of accomplishment such as pricing and profit. This, then, may be one of the obstacles in the planning and formation of library systems and networks. Library systems and networks are considered when broader library services are needed by the user. Properly planned systems and networks should improve the service at a reasonable cost.

One important difference is that special libraries file information rather than material, according to W. Ashworth in the "Handbook of Special Librarianship and Information Work."[41] He goes on to say that it is of great importance for the library staff to be willing and helpful in a special library. It must also be a place where requests are received and tenaciously pursued. Promotion of library services is accomplished through orientation, a well written guide, and library bulletins, with the best possible publicity being effective and willing service on a teamwork basis.

A parallel statement is made by Leon K. Albrecht in "Organization and Management of Information Processing Systems."[42] Albrecht says that the information systems organization operates in concert with all of the company's other organization entities and components. It must, therefore, be compatible with the company's environment and goals, and it should be utilized to the maximum.

In "Interpreting Services to the Library's Publics"[43] the science-oriented organization is described as depending on all types of specialized information pertinent to its needs. Not all this information may be in print and therefore may necessitate tapping individual expertise. Special libraries are established to develop the potential value of available facts. The library staff must be alert to user problems and strive to meet them courteously, accurately, efficiently, skillfully and creatively using knowledge of source material to extract pertinent information for transfer from media to minds. It is important to recognize the limits of budget and manpower in order not to "oversell" library service beyond what is practicable. A well developed list of communication techniques for public relations activities in special libraries is given at the end of this chapter on "Interpreting Services to the Library's Publics." This extensive list includes communications through display media, special services that broaden the reference service for the users and management and special communications with the library staff and professional colleagues.

Edwin B. Parker and Donald A. Dunn[44] in looking ahead to 1985 describe an information utility which could be made available to most U.S. homes. The system they discuss would provide better quality education and information to everyone. The unit costs of service are substantially less than present systems. A cooperative plan requires federal action as well as funds and support from the private sector. Technologies are now at the stage that could permit the creation of this information utility. This development could lead to an increased total expenditure on information services, just as the automobile led to increased expenditure on transportation. The costs are estimated to be low enough that society could afford to provide open and equal access to learning opportunities. General economic development may be stimulated by gains in economic productivity as a result of education.

In a more recent report[45] Professor Edwin B. Parker maintains that America is shifting from an industrial to an information society. This may seem an ancillary problem to the marketing of special libraries in industry. However, Parker cites a recent report from Japan which recommends that Japan spend well over $3 billion in five years to develop an "information society" capable of sustaining a 10% growth rate for the economy. This same study indicated that a *"laissez faire"* approach to information investment would support only a 7% growth. This is the same trend described by Parker in "Information Technology: Its Social Potential"[46] where he indicated development of the information utility could spark an economic surge comparable to the early development of railroads. Peter Drucker has argued that knowledge "has become the central capital, the cost center and the crucial resource of the economy." Parker observes that expenditures on information may constitute the most promising investment in improved economic activity, and further, that development of information utilities that permit economical, on-demand access to information services from most homes and offices may be the most promising road to national economic growth.

In trying to market the special library to industry one must try to understand organizational buying behavior. F.E. Webster, Jr. and

Yoram Wird in a "General Model for Understanding Organizational Buying Behavior"[47] try to describe this behavior.

In general, the authors do not find much reference to industrial buyers in the literature. However, their description of industrial buying is helpful in developing a marketing strategy for the industrial library. Industrial buying that takes place in an organization is influenced by budget, cost and profit considerations. Furthermore, organizational buying usually involves many people in the decision process with complex interactions among people and among individual and organizational goals. An industrial buyer behavior model can help the marketer identify the need for additional information, specify targets for marketing effort, identify kinds of information needed by various decision makers, and define the criteria they will use to make these decisions. This is an area in which in-depth study may be of considerable benefit.

American Libraries for February 1974 cites the following quotation by John Kenneth Galbraith under the headline "How to Sell a Library." "More than a decade ago, Galbraith was bemoaning the imbalance between the marketing of private goods and the marketing of public services. Thousands of dollars were spent to persuade the populace that it needed a new toothpaste, but who would sell the need for good schools and streets—and libraries?"[48] The answer in part is considered to be in the area of good public relations for all types of libraries. Janice Ladendorf is more emphatic in her article entitled "Breaking the User Barrier." She says that libraries are not really unique, but that they share many characteristics of the retail organization which has to sell its products. People usually have to be persuaded to use the library. Librarians should be applying market analysis or salesmanship to their customer relation problems. Every library, whether it is a public library or a special library, has its share of potential noncommitted users. Improving library utilization has got to be based on a continuing dialog. The success of any library can be measured by the amount of imagination and skill it shows in adapting to the unique needs of its client groups.[49]

Elizabeth Oakes in "Libraries and Sales Promotion"[50] states that people are influenced by advertising. For example, they support worthy causes because they have been persuaded to do so by promotional devices and techniques used by marketing specialists. Librarians are facing increasing competition and therefore must do a better job of promotion and marketing for their share of the budget or tax dollar. The small study reported by Mrs. Oakes showed the libraries using certain basic sales promotion devices and techniques usually received larger budgets than those not employing these techniques. Once again it was concluded that the library administration must become more imaginative and more skillful in its marketing and sales promotion practices in order to maintain or increase its budgets.

Almost 30 years ago Allen Kent reported on a survey of 100 special libraries of metallurgical companies. The thesis for the survey was the contribution that the library can make to the research effort. The conclusions indicated that there was a high positive correlation between the growth in earnings of companies in the metals industry and their degree of dependence on recorded information obtained through their libraries. The report was entitled "Literature Research as a Tool for Creative Thinking."[51] It concludes with the following statements:

> Creative work is inherent in library work. It behooves the metals librarian to be willing to devote effort to 'selling' the idea of the library as an important research tool within the organization.

The approaches cited in this paper concerning the "marketing" of special libraries are as varied as the special libraries that have tackled the problem. Many of the references said nothing at all about "marketing" although the entire thesis of their report was in fact a public relations activity presented in a serious and diplomatic "marketing" study. The special costs involved in setting up effective management information systems have been a major concern. MIS is a user-oriented information service which can affect operational changes for management. MIS is often developed for a specific information need for management. MIS failures occur through a lack of communication or a lack of understanding of management information needs.

A special library exists to supply information. The information supplied depends entirely on the users served. Both in Europe and in the United States, there are studies that indicate that a movement from an industrial society to an information or "knowledge" society may be in progress. The positive effect of this movement on economic growth has been indicated by several authors. Japan has assessed this movement and developed a policy for the country's economic benefit. The special library will be an integral part of this overall movement. Special libraries are increasingly participating in local, regional, and state-wide cooperative efforts. Special libraries in the private sector are being invited to join with the library service groups previously including only public funded libraries. As the national network idea becomes a reality for public funded institutions, it will also become a reality for special libraries that have been able to convince or "sell" their management on the value of participating in local public networks.

The "marketing" of the special library is essential. Marketing phraseology does not need to be used. However, communications and public relations activities are necessary. Some of the most emphatic "marketing" reports made strong effective statements on the user's need for information either to maintain the organizations leadership or to keep up with the competition. The advantage to be gained by thinking in marketing terms is that this is often the approach that is taken by management. Does the library provide the "right information" standard? Special libraries in industry are often asked to participate in company programs showing cost savings effected by changing clerical routines, adapting EDP systems for files and records, and participating in regional library networks and intracompany interlibrary loans. If we go a step further and ask ourselves if we are serving the users' needs with the right information, within the right time and at the best price, we will have the best "marketer" for the library and its services, a satisfied user.

"Marketing" of special libraries requires three elements: satisfied users, good communications, and a planned program of public relations activities. Special libraries must continue to practice marketing the "right information" standard but also do it at the most reasonable cost, and also do as much communication and public relations "selling" as time and staff permit. This is essential to their continued existence.

The National Science Foundation held a seminar in 1974 entitled "An Operational Experiment for the Marketing of Scientific and Technical Information Innovations."[52] The goals of this 2-year study include (a) extending the availability of scientific and technical information (STI) to a wider variety of user groups; (b) increasing the economically viable user base for STI and; (c) improving the receptivity to STI especially among user groups engaged in industrial R&D—the hypothesis to be tested is that improved information transfer leads to increased R&D productivity.

Industrial librarians were invited to this seminar and a composite of the "marketing" techniques of these libraries were discussed. Individual marketing techniques ranged from the very direct and personal visit of the librarian to every R&D staff member, to broader communication efforts through special bulletins, bibliographies, and announcements of services. The seminar endeavored to again identify "users" and answer the questions: What do the users want? When do they want the information? How do they want the information? What are the information requirements to support R&D in a company?

To develop the proper marketing of information, there is a need to know how the information is used. Can the information be supplied in better formats and also easier for the user to find? Is another system needed to supply STI for the user? As you can see, this was an exploratory session—the encouraging news is that the questions are being asked.

4. MODERN MARKETING ELEMENTS

Marketing success is achieved only if the product meets the standards or expectations of the consumer. Library success is achieved if the information meets the demands and needs of the user. Modern marketing elements include:

1. The product must meet the standards of the consumer.
2. The package is an effective marketing tool if it has both display effectiveness and favorable psychological connotations.
3. Advertising is most effective if it motivates one to buy without being conscious of the advertising.
4. People judge a product by its package (or a book by its cover).
5. A name, symbol, slogan or color may determine success or failure.
6. Symbols are more effective than words in motivating consumers because people are not aware that symbols affect their behavior.
7. Marketing must use words in correlation with symbols to communicate with consumers.
8. Creativity, originality and uniqueness have value in marketing communication if they motivate the consumer to buy.

By studying and adapting some of the above marketing elements for the promotion of the special library and its services, further success in marketing the special library could be achieved. Personnel do judge a library by its appearance; and if all the books look old and out of date, the judgment is made that the library is also out of date. It does not matter to the user that all the new books are being used. Appearance does matter. The latest edition of Beilstein closely resembles the earlier volumes published 50 years ago. The set looks old-fashioned and out of date except to the organic chemist. On the other hand, the old-fashioned and worn-out general encyclopedia set probably is out of date. If your library is furnished with cast-offs and make-do items, the casual visitor and the new employee are going to unconsciously rank the library as the last place to go for information. No manager worth his salt will stand for an office equipped with cast off furniture. It definitely conveys the wrong idea to his visitors. It is exactly the same with the library. And in a research establishment most of the researchers visit the library at least once. It should make a favorable impression as a working, useful and up-to-date library the first time. If it doesn't, there usually is not a second chance. The library is like a packaged product; if the "package" doesn't appeal to the user there is usually very little chance to get the user to come in and explore the contents.

5. SUGGESTED MARKETING RESEARCH STUDIES

In order to determine the status of industrial research libraries, a study based on a selection of libraries associated with the research laboratories listed in *Industrial Research Laboratories of the U.S.* might be helpful. The survey would endeavor to learn the location of the library in the organization; essentially to whom does the library management report. Budgets are always very sensitive questions, but it might be possible for the library to calculate the percent of the total industrial research budget that was spent for the library in one year. This figure may be three-tenths of one percent or maybe one percent of the overall budget. For more or less comparable figures, an example of what should be included in the total budget and the total library budget should be given. There would probably still be some libraries that could not respond to such a question, but perhaps there would be enough to give some idea of the trend of industrial library support. Other questions one might ask would include circulation statistics, publications and examples of the publications if possible, frequency of publication, coverage by house organs and internal plant newspapers and how often library feature stories were included. It might be possible to judge from that information the visibility of the library and if there was a correlation between library visibility in the media and library support.

With so many new computer-based services from which to pick, the package is important. It is important for the computer print-out that the user sees to be easy to use. It must identify the elements of the print-out and it must be printed evenly. There is a documented case where the letters on the line of a computer print-out were very uneven. All the information was there and it was the information that the user wanted. However, he cancelled the service and preferred to go without rather than try to read the uneven computer output. Here is a case where the quality of the information was excellent but the medium of communication so bad that it was not used for that reason alone. The elements of marketing are important in unconscious ways and affect decisions made by others about the library and library service. The packages that comprise the library and library services should be studied for the most effective means of communication.

6. CONCLUSION

Data from these studies would give valuable information to libraries concerning their image. An awareness of marketing elements and how they influence users and decision makers could be utilized to strengthen and improve the position and role of the industrial special library. Many of the studies examined for this review were the result of a need for a library or expanded library services. The need for the industrial library exists but its success or failure relies on the utilization of good marketing techniques as well as providing excellent service. To help librarians understand and utilize the concepts of modern marketing techniques, continuing education seminars could be scheduled on this subject. Special librarians have excellent records for providing services. Proper marketing can aid immeasurably in the extension of these services by utilizing marketing support and packaging techniques for the libraries and their products.

REFERENCES

1. Clemans, M. Containing the information explosion. *Optima*. 18(4): 186-189; 1968 December.

2. Herner, Saul; Heatwole, M. K. The planning of libraries for military research establishments. *Science*. 114:57-59; 1973 July 20.

3. Blaxter, K. L.; Blaxter, Mildred L. The individual and the information problem. *Nature*. 246: 335-339; 1973 December 7.

4. Colinese, P. E. Keeping in touch with the needs of information users. Institute of Information Scientists. *Proceedings of Second Conference.* Held at Jesus College, Oxford, on July 11-13, 1966. London: The Institute; 1968: 63-68.

5. Allen, Thomas J.; Cohen, Stephen I. Information flow in research and development laboratories. *Administrative Science Quarterly.* 14: 12-19; 1969.

6. Uvarov, E. G. Starting a small industrial library. *Research.* 5: 510-514; 1952.

7. *Information units in small plants.* New York: United Nations; 1973. 48 p.

8. Anderla, George. A challenge for governments and society. *OECD Observer.* 1: 27-32; 1973 April.

9. Veazie, Walter H., Jr.; Connolly, Thomas F. *The marketing of information analysis products and services.* Washington, D.C.: ERIC Clearinghouse on Library and Information Sciences; 1971.

10. Sviridov, F. A. International cooperation among information specialists: the work of FID. *ASLIB Proceedings.* 22(8): 377-385; 1970.

11. Scheffler, F. L.; March, J. F. *User appraisal and cost analysis of the Aerospace Materials Information Center.* Dayton, OH: Dayton University Research Institute; 1970 March; AD 707143; AFML TR 70-27.

12. Vann, James O. *Defense Documentation Center* (DDC) for scientific and technical information. Journal of Chemical Documentation. 3(4): 220-222; 1963 Oct. Alexandria, VA: the Center; n.d.

13. North American Aviation, Inc. *Analysis of scientific and technical information requirements of FAA.* Contract No. FA64WA5213. Downey, CA; 1965 Oct 25; SID-65-1397; AD 651926. 160 p.

14a. Committee on Scientific and Technical Information. *Progress in scientific and technical communications. Annual Report No. 6.* Washington: Federal Council for Science and Technology; 1969; PB 186400. 99 p.

14b. Committee on Scientific and Technical Information. *Progress in scientific and technical communications. Annual Report No. 7.* Washington; Federal Council for Science and Technology; 1970; PB 193386. 166p.

15. Committee on Scientific and Technical Information. *Scientific and technical communication: a pressing national problem and recommendations for its solution.* Washington, DC: National Academy of Sciences; 1969. (SATCOM report).

16. Chorafas, Dimitris N. *The knowledge revolution; an analysis of the international brain market.* New York: McGraw-Hill; 1970. 142p.

17. Vernimb, Carl O.; Steven, Gunter. ENDS-European Nuclear Documentation Service. *Nuclear Engineering and Design.* 25(3): 325-333; 1973 Aug.

18. Hillier, James. Measuring the value of information services. *Journal of Chemical Documentation.* 2:31-34; 1962.

19. Merritt, C. Allen; Nelson, Paul J. *The engineer scientist and an information retrieval system.* Yorktown Heights, NY: IBM Technical Information Retrieval Center/ Thomas J. Watson Research Center; 1965 August 1; ITIRC-003 technical report.

20. Knox, William T. Systems for technological information transfer. *Science.* 181:415-419; 1973 August 3.

21. Ferguson, Elizabeth. Public relations in special libraries. *In:* Angoff, Allen. *Public relations for libraries.* Westport, CT: Greenwood Press; 1973: 179-198.

22. Armstrong, Alan. For good decisions you need the facts. *Financial Times.* :8-9, 1973 December 17.

23. Licklider, J. C. R. *Libraries of the future.* Cambridge, MA: M.I.T. Press; 1965.

24. North Atlantic Treaty Organization. Advisory Group on for Aerospace Research and Development. Scientific and technical information. Why? Which? Where? and How? By H.A. Stolk. Neuilly-sur-Seine: NATO; 1971; AGARD-LS-44-71; Lecture series no. 44.

25. Luhn, H. P. A business intelligence system. *IBM Journal of Research and Development.* 2: 314-319; 1958 Oct.

26. Kashyap, R. N. Management information systems for corporate planning and control. *Long Range Planning.* :25-31; 1972 June.

27. Mintzberg, Henry. Myths of MIS. *California Management Review.* 15: 92-97; 1972.

28. Stout, Larry D. The problems of management information systems: why they fail. *GAO Review.* :40-45; 1972 Fall.

29. Gessford, John E. Management information systems development, Part I. *Managerial Planning.* 21: 15-29; 1973 January/February.

30. Gessford, John E. Management information systems development, Part II. *Managerial Planning.* 21: 1-18; 1973 March/April.

31. Hax, Arnoldo C. Planning a management information system for a distributing and manufacturing company. *Sloan Management Review.* 14: 85-89; 1973 Spring.

32. Long, Nicholas. Information and referral services: a short history and some recommendations. *Social Service Review.* 47: 49-62; 1973 March.

33. Gibson, Lawrence D. et al. An evolutionary approach to marketing information systems. *Journal of Marketing.* 37: 2-6; 1973 April.

34. Sloane, Leslie. Managing information. I. *Bankers Magazine.* (London). 215 (1549): 145-149; 1973 April.

35. King, William R.; Cleland, David I. Decision and information systems for strategic planning. *Business Horizons.* 16: 2936; 1973 April.

36. Reif, William E.; Monczka, Robert M. A high level independent location for the information systems department. *Arizona Business Bulletin.* 20: 17-24; 1973 May.

37. Burgstaller, H. A.; Forsyth, John D. The key-result approach to designing management information systems. *Management Adviser.* 10(3): 19-25; 1973 May/June.

38. *Conference on the Communication of Scientific and Technical Knowledge to Industry.* Held in Stockholm on October 7-9; 1963, Paris: Organization for Economic Cooperation and Development, 1965.

39. Jahoda, Gerald. Special libraries and information centres in industry in the United States. *UNESCO Bulletin for Libraries.* 17: 70-76; 1963 March/April.

40. Nelson, Charles A.; Nelson, Anne H. Library systems and networks. *Management Controls.* :71-75; 1973 March.

41. Ashworth, Wilfred. *Handbook of special librarianship and information work.* 3d ed. London; Aslib: 1967.

42. Albrecht, Leon K. *Organization and management of information processing systems.* New York: Macmillan; 1973.

43. Strauss, Lucille J.; Shreve, Brown, Alberta L. *Scientific and technical libraries: their organization and administration.* 2d ed. New York: Becker and Hayes; 1972: Chap. 12.

44. Parker, Edwin B.; Dunn, Donald A. Information technology: its social potential. *Science.* 176: 1392-1399; 1982.

45. Posner, Edwin B.; Shifting from industrial to information society (abstract). *GDP's US/R&D.* 390: 1973 Dec.

46. Parker and Dunn, *op. cit.*

47. Webster, F. E.; Wird, Yoram. A general model for understanding organizational buying behavior. *Journal of Marketing.* 36: 12-19; 1972.

48. How tc sell a library (book review). *American Libraries.* 5(2): 81; 1974 Feb.

49. Ladendorf, Janice. Breaking the user barrier. *RQ.* 11(4): 337-339; 1972 Summer.

50. Oakes, Elizabeth. Libraries and sales promotion. *California Librarian.* 33(3): 155-163; 1972 July.

51. Kent, Allen. *Creative thinking in the library.* Cleveland: Center for Documentation and Communication Research, Western Reserve University; 1956 June 4; PB 166441. (Presented at SLA Conference held in Pittsburgh, on June 4, 1956.)

52. *Seminar on an Operational Experiment for the Marketing of Scientific and Technical Information Innovations.* Held in 1974. Washington, National Science Foundation; 1974.

SCI-TECH COLLECTIONS

Tony Stankus, Editor

Steve Hunter's contribution is one of the more comfortably literate in this continuing series. This is fortunate for two reasons. First, only a small but much appreciated portion of our readership serves a clinical clientele, and are thereby familiar with the terminology and literature. Second, mounting social and political concern coupled with amazingly little fundamental understanding will soon present funding opportunities for the more basic science-oriented clientele that a majority of us serve. With the goal of keeping our readers one step ahead, I'm happy to present this paper. In closing I again invite potential contributors to submit their ideas for papers along with a statement of their qualifications. While outlines are often readily approved, each contribution is subject to editing and the editors must, of course, reserve the option of rejection.

Alzheimer's Disease: A Guide to Information Services

R. Stephen Hunter

INTRODUCTION

Alzheimer's disease is a degenerative process in the brain that produces progressive intellectual impairment and ultimately leads to death. It has been deemed by some to be the disease of the century.[1] Lately, it has been the subject of television documentaries and numerous popular magazine and newspaper articles, including the cover story of a recent issue of *Newsweek*. It has captured the attention of the media in ways reminiscent of AIDS or Toxic Shock Syndrome, and consequently is becoming a household word. Indeed, because memory loss is an important clinical feature of the disorder, it is no wonder that "Alzheimer's disease" is creeping into the popular lexicon, gradually usurping "senility" as the word of choice when one forgets an appointment or the name of a passing acquaintance.

Why all the sudden interest in a disease first described in 1907? There are several reasons which will be addressed through the course of this paper, but perhaps the most significant reason is the "greying" of America. Alzheimer's disease is predominantly a disease of the aged, and there are now twenty-seven million Americans over the age of sixty-five, with this figure projected to increase to forty-five million by the end of the century. About three million of these twenty-seven, or seven percent, are afflicted with Alzheimer's. Close to eighty percent of the nursing home population suffers from the disorder, and between 100,000 and 120,000 people die from it annually, making it the fourth leading cause of death among the elderly after heart disease, cancer and stroke.[2] Add

R. Stephen Hunter is Library Director, Worcester Memorial Hospital, 119 Belmont Street, Worcester, MA 01605. He holds a BA from Worcester State College and an MLS from the University of Rhode Island at Kingston.

to these numbers the family members who must cope with spouses or parents with Alzheimer's and it is easy to realize the impact this disease has on the population.

The purpose of this article is to provide a guide to the literature of Alzheimer's disease. Because Alzheimer's is a specific medical entity, the biomedical and health sciences literature will be emphasized. There is a growing body of literature intended for lay readership, and while not comprehensively treated, significant and important sources of lay information will be noted. Since Alzheimer's is a single entity, the scope of the literature is not as encompassing as some subjects recently reviewed in this column. Monograph titles mentioned within the text are given full bibliographic citations in the Appendix.

THE DEMENTIAS

Alzheimer's disease is a degenerative, progressive organic brain syndrome, and is usually grouped within the broader category of dementia. "Dementia" corresponds to the Latin word dementatus—that is, out of one's mind, mad, crazed.[3] Until the end of the 18th century, dementia was usually used to mean insanity, and is still used that way to some extent in common parlance.

Dementia, however, is the medical term for a group of symptoms and not a specific disease entity. It describes a global decline in intellectual ability sufficiently severe to interfere with a person's daily functioning, and occurs in a person who is awake and alert (not drowsy or intoxicated).[4] This decline in intellectual functioning means a loss of several kinds of mental processes that include memory, mathematical ability, vocabulary, abstract thinking, judgement, speaking, or coordination, and may include personality changes.

The symptoms of dementia can be caused by many diseases. Some are treatable; some are not. Dementia can be reversed or stopped in some while in others it cannot be changed, which is true of Alzheimer's. Alzheimer's disease is one of a group of degenerative diseases with generally unknown causes which also include Friedreich's ataxia, Huntington's disease, Parkinson's disease, Pick's disease, Progressive supranuclear palsy, and Wilson's disease.[5] Of these, Alzheimer's is by far the more prevalent.

Historically, dementia was sometimes thought to be the result of an indwelling spirit, but with the advent of the 19th century was recognized as an illness. To the medical scientist, a disease is a clinical and pathological entity: a characteristic and usually progressive set of changes in the appearance and function of the body and in the gross structure of affected organs or tissues. In the 1890s Kraepelin recognized that among the group of "insanities", insanity in the aged was one of the few psychiatric syndromes accompanied by gross brain changes. In his postmortem studies, half of the group of these "insane" aged had cerebral infarcts, which he attributed to arteriosclerosis or hardening of the arteries, and the other half displayed atrophy of the cerebral cortex without significant arteriosclerosis.[6]

Kraepelin's work was continued by the German neurologist Alois Alzheimer, who further delineated the pathologic changes in the brain that essentially are the same used today for a definitive diagnosis of Alzheimer's disease. These changes will be discussed in more detail later in this paper.

Until recent years, the disease named for Alzheimer was considered to be a specifically "presenile" dementia, stemming from Alzheimer's description of the disease in a woman of fifty-one.[7] Elderly people with comparable symptoms were said to be suffering from hardening of the arteries. "Presenile" was the term used to distinguish the entity from dementia of the "senile" type, i.e., occurring in persons over the age of sixty.

Medical scientists now agree, however, that the dementia that occurs in the elderly is the same as the "presenile" condition. Consequently, Alzheimer's disease occurring in persons past the age of sixty is commonly defined as senile dementia of the Alzheimer type in order to delineate by age of onset.[8]

CLINICAL ENTITY

Alzheimer's disease can be defined as a chronic cognitive dysfunction which, as mentioned earlier, is also progressive and age-related. The current belief is that cognitive dysfunction is not an inevitable concomitant of old age, since many elderly apparently do not experience significant memory impairment. Alzheimer's is accompanied by cognitive and behavioral symptoms, the precise nature of which depends on the stage of the disease.

Alzheimer's is usually classified into three stages.[9] In the first stage, memory loss and behavior or personality changes such as lack of spontaneity or loss of sense of humor are characteristic. The second stage is characterized by progressive memory loss, impaired speech, agnosia, impaired coordination, wandering, and repetitive movements such as tapping, pacing, lip-licking or chewing. In the third or terminal stage, victims often do not eat, become emaciated, are unable to communicate, may be incontinent, and may have grand mal seizures. Death usually results from conditions common to bed-ridden, comatose patients, such as pneumonia.[10] Hardly ever is Alzheimer's disease officially cited as the cause of death.

DIAGNOSIS

Alzheimer's is a difficult entity to diagnose. A definitive diagnosis of Alzheimer's disease usually can only be made after a process of eliminating many other diseases or syndromes that may present with similar symptomatology. Such things as toxic conditions, nutritional disorders, certain infections, endocrine disorders, and brain tumor produce similar symptoms, and are classified as secondary dementia.[11] Multi-infarct dementia, repeated strokes within the brain which destroy small areas of the brain, and depression must also be ruled out. Most cases of secondary dementia can be reversed if they are identified and treated early in the course of the illness.

There are several diagnostic tests that can be administered to rule out secondary causes of dementia, but it is only a brain biopsy that can make a positive diagnosis of Alzheimer's.[12] A brain biopsy is done by removing a piece of skull bone and taking out a small sample of brain tissue. This test is not done routinely at present because no effective treatment is available even if a positive diagnosis is made.[13]

When all other possibilities are eliminated, only then will a diagnosis of Alzheimer's be established, and then it is usually qualified as very probable. Kerzner has authored an excellent review article on the diagnosis of Alzheimer's from the clinician's point of view. (See author's bibliography.)

INCIDENCE

The prevalence rate for Alzheimer's disease has been reported to range from 0.4 percent to 18 percent,[14] certainly a wide discrepancy. The figure of 7 percent in the population sixty and over is generally accepted to be a reliable number.[15] Some studies have indicated a higher prevalence in women,[16] but since there are more women than men who live past the age of sixty-five, the data may be biased. Other studies have shown an equal prevalence among men and women.[17]

PATHOLOGY

Alzheimer's disease is characterized by microscopic physical changes in the brain. Postmortem studies have consistently revealed some major pathological changes that remain enigmatic to researchers. The weight of the brain is not significantly different between Alzheimer's victims and "normal" subjects of the same age. Some show marked atrophy of the cerebral cortex, while others show little or no atrophy at all. Thus, while atrophy is a prominent feature in Alzheimer's, a large number of cases have no evidence of atrophy.[18]

Microscopically, the major pathological changes or lesions are neurofibrillary tangles and senile plaques. Neurofibrillary tangles refer to accumulations of twisted brain tissue fillaments. Senile plaques are aggregates of cellular debris and amyloid protein. In addition, there is a significant loss of neurons in the cortex, basal nucleus and in the region next to the hippocampus, with consequent reduction in the amount of neurotransmitters or chemical messengers.[19]

ETIOLOGY

The etiologies or causes of Alzheimer's disease are unknown. Researchers are currently focusing on six hypothetical models of pathogenesis. Each of these models is the subject of extensive coverage in the journal literature, and will only be briefly mentioned here. An excellent overview of these models is contained in Wurtman's *Scientific American* article.[20]

All of these models have strong evidence to support research, but as yet nothing is conclusive. The models are genetic, the abnormal protein, the infectious agent, the toxin, the blood flow, and the acetylcholine. Wurtman also mentions an additional model which he calls the "elephant" model, coined for the parable of the blind men who all describe an elephant differently depending on the part of the elephant's anatomy they happen to touch. All pose descriptions of the animal, but they do not perceive its essential elephantness. He points out that past investigators had hypothesized that cancer was caused by virus, diet, impaired or faulty genetic make-up, environmental toxins, radiation or immune system deficiencies. Because all of these hypotheses turned out to be correct, research is now centering on what may be the essential "elephantness" of cancer. Perhaps the same will be true of Alzheimer's disease.

PROGNOSIS AND MANAGEMENT

Alzheimer's disease is incurable, and its progress inexorable. The average length of time for a person to pass through its three stages is six to eight years, but can take up to twenty. The factors influencing the progression of a patient from the forgetfulness phase to the impaired confusional stage and then to the demented stage are unknown. Because its onset is so insidious and the signs and symptoms vary widely, diagnosis is not usually made until the later stages of the disease. Some progress rapidly from the first to the third stage, others may show an equivalent change over the course of a decade, so it is very difficult to generalize on prognosis. The life expectancy of patients in the third stage is definitely decreased because these people lose the ability to care for themselves.

Since the etiology of Alzheimer's remains unknown, therapy aimed at arresting or curing the disease has been unsuccessful. There is no medication available that will predictably improve cerebral function. Therefore, therapy must have the alteration of symptoms as its goal.[21] Symptoms depend in large part on the patient's social situation. If he has relatives who are willing and able to support his activities, he may remain relatively comfortably in the home environment for years. If he has no support from his family or community, he will become symptomatic much earlier and will require custodial care early in the illness. Firm recommendations cannot be made for therapy that will include all patients; each person and each situation require individual assessment.

LITERATURE FOR A GENERAL AUDIENCE

As mentioned earlier in this paper, the public's awareness of Alzheimer's disease is on the rise as the subject becomes trendy in the popular press. In addition to readers of general interest, there is also a great need for family members of Alzheimer's victims to become better informed of the condition.

Mace and Rabins' *The 36-Hour Day* provides an excellent overview of all aspects of the disease. As its subtitle suggests, it is intended as a guide for family members, but can be considered must reading for anyone beginning to obtain information about Alzheimer's disease, whether a lay or professional person. It also contains an excellent bibliography geared to the general reader.

Another excellent guide for families is Zarit et al.'s *Caring for the Patient with Alzheimer's Disease*, recently published by the New York University Press. Reisberg's *Alzheimer's Disease* is another recent monograph that will be enlightening to the lay reader as well as the professional. Heston's book is yet another of the same scope.

There are many articles in popular magazines that provide good overviews of the disease. These can be extensive or concise, depending on the intended market. The *Newsweek* article mentioned earlier provides a concise, highly readable treatment of the disease and was well researched by its authors. Wurtman's *Scientific American* article, while aimed at the more sophisticated reader, offers a comprehensive treatment of the current research into the causes of Alzheimer's. *Magazine Index* and *Reader's Guide to Periodical Literature* can be searched for other material using the subject heading Alzheimer's Disease.

The Alzheimer's Disease and Related Disorders Association, Inc., 360 North Michigan Avenue, Chicago, IL, 60601, is a good source of reliable information for lay readers and family members, and has published numerous pamphlets and short monographs. This organization also has local chapters that will provide these materials.

BOOKS FOR THE PROFESSIONAL

Seeking information on Alzheimer's disease in book or monograph format can be discouraging. In textbooks, information concerning Alzheimer's tends to be scattered and sparse. Authoritative texts in neurology, psychiatry and gerontology, where one would expect to find good overviews of the disease, all offer disappointing coverage. Such respected works as Kaplan's *Comprehensive Textbook of Psychiatry*, Merritt's *Textbook of Neurology*, or Reichel's *Clinical Aspects of Aging* devote only one or two pages to the subject. No doubt future editions will offer more as research into Alzheimer's disease becomes more conclusive.

Some excellent monographs have been published in recent years, however, indicative of the progress in research. Of the forty-nine monographs on the subject catalogued by the National Library of Medicine, thirty-one have an imprint within the last five years, and sixteen of these were published in 1983 or 1984. Several of these, titles by Mace, Zarit, Heston and Reisburg, have been mentioned previously as being of some interest to general readers. For extensive coverage of recent developments, *Biological Aspects of Alzheimer's Disease*, edited by Katzman and published by Cold Spring Harbor Laboratory, is state-of-the-art. Another fine treatment of current research can be found in *Alzheimer's Disease: Advances in Basic Research and Therapies*, edited by Wurtman and recently published by the Center for Brain Sciences and Metabolism, which contains the proceedings of the third meeting of the International Study Group on the treatment of Memory Disorders Associated with Aging, which convened in Zurich in January of 1984. Kelly's *Alzheimer's Disease and Related Disorders* is another recent publication of the same genre, and has information that can be useful to the clinician as well as the researcher.

Raven Press, a publisher of many neurosciences titles, offers a compilation of papers related to current research in *Alzheimer's Disease: A Report of Progress in Research.* Another title published by Raven Press, actually an irregular serial, is *The Dementias*, edited by Mayeux, and Volume 38 of their *Advances in Neurology* series. This volume contains many papers related to Alzheimer's, and can be considered an excellent review.

The Select Committee on Aging of the U.S. House of Representatives published a series of reports of hearings before the House in 1983. These reports tend to verbosity and may be of only passing interest to the scientist, but provide some important insight into the recent attention the disease is getting and into the role the government has taken in funding Alzheimer's research.

Appropriate subject headings for books catalogued with the National Library of Medicine scheme are Alzheimer's Disease or Dementia, Presenile. The NLM only started using Alzheimer's Disease in 1984; previously, works were classed as Dementia, Presenile. The Library of Congress has been using Alzheimer's Disease as a subject heading since 1982; earlier imprints were classed as Presenile Dementia.

JOURNAL LITERATURE

It has become axiomatic to state that it is the journal literature that contains the bulk of scientific thought and research, and is the print format most frequently utilized by scientists seeking information. Nevertheless, the statement holds true for the literature of Alzheimer's disease. There are no journals currently being published that are exclusively devoted to research in Alzheimer's disease, although one can easily speculate that this may not remain true for much longer given the interest the subject has been generating. Given the scattered nature of Alzheimer's literature, a peer-reviewed, interdisciplinary journal would no doubt receive some enthusiasm from health science professionals.

Before taking a look at the recent expansion of citations relevant to Alzheimer's in the literature, mention should be made of some excellent recent review articles. The *Index Medicus* includes a Bibliography of Medical Reviews in each monthly issue, and is published in Volume 2 of its annual cumulation. The *Bibliography of Medical Reviews* is an excellent source for reviews of the literature in subject areas of interest to biomedical scientists and practitioners, and is an appropriate place to begin searching the literature. In the preparation of this article, the author found several review articles particularly comprehensive. Reviews by Rathman, Kerzner, Coyle and Schneck have been cited earlier, and are fully described at the end of the article.

Interest in Alzheimer's is multidisciplinary, and articles tend to be diffused throughout the periodical literature of major disciplines such as Neurology, Gerontology, Psychiatry and Internal Medicine. To get a general idea of the journals by their field as well as specific titles that contain a high percentage of Alzheimer's related articles, a search of the current MEDLINE file (1983-), accessed through the National Library of Medicine, was accomplished. An analysis of the results of the search by broad subject is contained in Table 1. Table 2 analyzes specific titles that contain significant numbers of "hits".

INDICES AND DATABASES

The most reliable access to the literature in print format are the major biomedical indices *Index Medicus, Biological Abstracts, Science Citation Index* and *Chemical Abstracts*, with the *Index*

Table 1

Number of postings in current MEDLINE file as of February 1985: 562

Subject	Number of Postings	Percentage of Total
General Medicine (M)*	90	16.0
Neurology (N)	229	41.0
Gerontology (G)	64	11.0
Psychiatry (P)	61	11.0
Other (Pharmacology, Genetics Biochemistry, etc.)	118	21.0

*Included in this category are journals with a broad scope of interest to the medical community, e.g., New England Journal of Medicine, JAMA, Lancet, Annals of Internal Medicine, American Family Physician

Table 2

Title (Category)	Number of Citations	Percentage of Total Citations	Percentage of Category
Annals of Neurology (N)	25	4.4	10.9
Archives of Neurology (N)	14	2.4	6.1
Brain Research (N)	14	2.4	6.1
American Journal of Psychiatry (P)	8	1.4	13.1
Journal of Neurology, Neurosurgery, and Psychiatry (N)	15	2.6	6.5
Neurology (N)	16	2.8	7.0
Journal of Neuroscience (N)	14	2.4	6.1
Journal of the American Geriatric Assoc. (G)	17	3.0	26.5
Journal of Clinical Psychiatry (P)	6	1.0	9.8
New England Journal of Medicine (M)	6	1.0	6.6
Lancet	10	1.8	11.0

Medicus being the better known and more ubiquitous. Alzheimer's Disease is the most appropriate subject heading to search the *Index Medicus* since January 1984. Prior to that date, the subject heading used for papers related to Alzheimer's was Dementia, Presenile. (In the MEDLINE databases, however, Alzheimer's Disease can be used to search the literature back to 1975; for the literature back to 1966, Dementia, Presenile and the truncated textword Alzheimer is perhaps the most effective strategy.) These subject headings are relevant for searching the other major indices.

In order to obtain an idea as to the coverage of the literature in the various databases related to biomedicine, a free text search was conducted on the databases MEDLINE, BIOSIS, EXMED, CAS, LIFE SCIENCES, and SCISEARCH. These databases consistently have the highest number of postings. Depending on the nature of information desired, other databases such as Magazine Index and SOCIAL SCISEARCH should not be overlooked. The results of this search are listed in Table 3.

In order to stress the recent growth of relevant citations in the literature, a search was also done of the literature for the last five years (1980-1984). Table 4 summarizes the results of this search, and indicates the percentage of the total postings in the corresponding database. These results bear testimony to the current awareness and focus of research in Alzheimer's disease.

Table 3

Databases Related to the Literature of Alzheimer's Disease

DATABASE (Years Covered	Postings
MEDLINE (1966-)	1646
BIOSIS (1969-)	1511
EXMED (1975-)	1394
SCISEARCH (1974-)	925
CAS (1974-)	287
LIFE SCIENCES (1978-)	98

Table 4

Database	Last 5 Years	Percentage of Total Postings in Database
MEDLINE	1248	79
BIOSIS	1080	71
EXMED	786	56
SCISEARCH	705	76
CAS	251	81
LIFE SCIENCES	90	92

CONCLUSION

Research into the causes and treatment of Alzheimer's disease is ongoing, and there is hope that medical science is beginning to unravel the enigma of the illness. Since 1976, federal research spending on Alzheimer's has increased from 4 million to 37.1 million in 1984, and the literature of Alzheimer's has enjoyed a comparable increase. Refined investigative techniques have provided provocative clues, and these discoveries raise the possibility that treatments will be found. No longer thought to be an inevitable part of aging, scientists compare the current status of Alzheimer's to the challenge they faced with heart disease thirty years ago.[22] Advances in research are finding their way into print at an exponential rate, and this condition is likely to remain true for the foreseeable future.

REFERENCES

1. Clark, Matt. A slow death of the mind. *Newsweek*. 56-62; 1984 December 3.
2. Toseland, R.W. Alzheimer's disease and related disorders: assessment and intervention. *Health and Social Work*. 9(3): 212-26; 1984 Summer.
3. Kaplan, Harold I.; Freedman, Alfred M.; Sadock, Benjamin J. *Comprehensive textbook of psychiatry*. 3d ed. Baltimore: Williams and Wilkins; 1980.
4. Schneck, Michael K.; Reisberg, Barry; Ferris, Steven H. An overview of current concepts in Alzheimer's disease. *American Journal of Psychiatry*. 139(2): 165-73; 1982 February.
5. Mace, Nancy L.; Rabins, Peter V. *The 36-hour day*. Baltimore: Johns Hopkins University Press, 1981.
6. Torack, R.M. *The pathologic physiology of dementia*. New York: Springer-Verlag; 1978.
7. Ibid.
8. Mace. Op. Cit.
9. Gwyther, Lisa P.; Phillips, Linda R. Care for the caregivers. *Journal of Gerontological Nursing*. 9(2): 93-95, 110, 116; 1983 February.
10. Schneck. Op. Cit.
11. Kerzner, Lawrence J. Diagnosis and treatment of Alzheimer's disease. *Advances in Internal Medicine*. 29: 447-70; 1984.
12. Wurtman, Richard J. Alzheimer's disease. *Scientific American*. 252(1): 62-75; 1985 January.
13. Rathmann, K.L.; Conner, C.S. Alzheimer's disease: clinical features, pathogenesis, and treatment. *Drug Intelligence and Clinical Pharmacy*. 18(9): 684-91; 1984 September.
14. Schneck. Op. Cit.
15. Roth, Michael. Epidemiological studies. In: Katzman, Michael, ed. *Alzheimer's disease, senile dementia and related disorders*. New York: Raven Press; 1978.
16. Mace. Op. Cit.
17. Roth. Op. Cit.
18. Schneck. Op. Cit.

19. Wurtman. Op. Cit.
20. Coyle, J.T.; Price, D.L.; DeLong, M.R. Alzheimer's disease: a disorder of cortical cholinergic innervation. *Science*. 219(4589): 1184-90; 1983 March.
21. Kaplan. Op. Cit.
22. Clark. Op. Cit.

APPENDIX: MONOGRAPHS DISCUSSED

Corkin, Suzanne. *Alzheimer's disease, a report of progress in research.* New York: Raven Press; 1982.
Heston, Leonard L.; White, June L. *Dementia, a practical guide to Alzheimer's disease and related illnesses.* New York: Freeman; 1983.
Katzman, Robert, ed. *Biological aspects of Alzheimer's disease.* Cold Spring Harbor, NY: Cold Spring Harbor Laboratory; 1983.
Kelly, William E. ed. *Alzheimer's disease and related disorders*. Springfield, IL: Thomas; 1984.
Mace, Nancy L.; Rabins, Peter V. *The 36-hour day*. Baltimore: Johns Hopkins University Press; 1981.
Mayeux, Richard; Rosen, Wilma G. The dementias. *Advances in Neurology*. v.38. New York: Raven Press; 1983.
Reisberg, Barry, ed. *Alzheimer's disease.* New York: Free Press; 1983.
Wurtman, Richard J., ed. *Alzheimer's disease, advances in basic research and therapies.* Cambridge, MA: Center for Brain Sciences and Metabolism Charitable Trust; 1984.
Zarit, Steven H.; Orr, Nancy K.; Zarit, Judy M. *Caring for the patient with Alzheimer's disease.* New York: New York University Press; 1985.

NEW REFERENCE WORKS IN SCIENCE AND TECHNOLOGY

Robert G. Krupp, Editor

Reviewers for this column are: Carmela Carbone (CC), Engineering Societies Library, New York, NY; Kerry Kresse (KLK), University of Kentucky, Lexington, KY; Robert G. Krupp (RGK), Maplewood, NJ; Ellis Mount (EM), Columbia University, New York, NY; Barbara List (BL), Columbia University, New York, NY; and David A. Tyckoson (DAT), Iowa State University, Ames, IA.

BIOMEDICAL SCIENCES

(A) Catalog of the Diptera of America north of Mexico. Prepared cooperatively by specialists on the various groups of Diptera under the direction of Alan Stone and others, Agricultural Research Service, United States Department of Agriculture. Washington: Smithsonian Inst. Pr.; 1983. 1696p. $37.50. ISBN 0-87474-890-9. (Agriculture Handbook; no. 276.) (Reprint. Originally published: Washington: U.S. G.P.O.; 1965.)

With this reprint once again a catalog of the true flies found in North America north of Mexico becomes available for purchase. The four main purposes for compiling this work are to list all published names with reference to the original publication of each, to indicate valid and synonymous names, to provide a classification, and to indicate distribution. The more than 25,000 entries cover two-winged flies including midges, gnats, and mosquitoes, in addition to all other recognized flies. Geographically, the scope includes Diptera found in Greenland, Bermuda, and the California Channel Islands as well as other regions north of Mexico. An important section provides nearly 4,800 references to selected works of North American Diptera used in the catalog or in annotations within the bibliography itself. This volume certainly should be included in any collection of systematic zoology or entomology. (BL)

Dictionary of American medical biography. Edited by Martin Kaufman and others. Westport, CT: Greenwood Pr.; 1984. 2 volumes. $95.00. ISBN 0-313-21378-X (lib. bdg.) set.

This is a source of biographical information on people important to American medicine from the 17th century up through the mid-1970's. It does not include those who have died after December 1976. A special attempt has been made to identify blacks and women who had not received attention in earlier works. In addition, biochemists, medical educators, and hospital administrators of importance are within the scope of this work. People on the fringes of medicine such as health faddists, patent medicine manufacturers, and unorthodox practitioners can also be found. Where possible, entries give date and place of birth, date and place of death, occupation, area of specialization, parents' names and occupations, marital status, names of spouses, number of children, career including dates and positions, and finally, contributions to American medicine and writings. Five appendices organize the people by date of birth, place of birth, state where prominent, specialty or occupations, and medical or graduate college. (BL)

Directory of food and nutrition information services and resources. Edited by Robyn C. Frank. Phoenix, AZ: Oryx Press; 1984. 287p. $85.68 (pap.). ISBN 0-89774-078-5.

Ms. Frank, director of the U.S. Department of Agriculture's Food and Nutrition Information Center, has put a tremendous amount of effort into this directory. Aimed at librarians, educators, and health professionals, the *Directory* identifies a wide range of information sources for food and nutrition. It is divided into nine sections, and the first three (organizations, databases, and microcomputer software) were compiled from mailed questionnaires. The remaining chapters (journals and newsletters, indexes and abstracts, producers of food and nutrition materials, reference works, governmental and private agencies, and tables of nutrient values for common foodstuffs) were compiled by the editor. Access to these many items is provided by a variety of indexes. An attempt was made to provide reliable resources, and a series of cited guidelines was followed toward that end. It is an ambitious project and a worthwhile purchase. Recommended for academic libraries, special libraries, and large public libraries. (KLK)

(The) Encyclopedia of alcoholism. By Robert O'Brien and Morris Chafetz. New York: Facts on File; 1982. 378p. $40.00. ISBN 0-87196-623-9.

More than 500 entries arranged in one alphabetic sequence offer concise, dictionary-like definitions. The treatment is broad and covers the substance alcohol; social institutions, customs and socioeconomic aspects of alcoholism; and physical and psychological manifestations of alcoholism. The entries range in length from one sentence to several pages. Major entries are followed by a short list of references. There are two appendices, the first of which consists of 43 tables organizing hard to find facts, plus maps and figures. The second appendix lists selected sources of information covering organizations (excluding treatment centers) and English-language journals and periodicals. Complete addresses are given for each. A twenty-four page bibliography on the subject will be useful for the professional and lay person alike. (BL)

(The) Encyclopedia of mammals. Edited by David Macdonald. New York: Facts on File; 1984. 895p. $45.00. ISBN 0-87196-871-1.

This colorful encyclopedia provides articles on the life, habits, and physiological features of over 4,000 species of mammals. Chapters are grouped into six main categories—carnivores, sea mammals, primates, large herbivores, small herbivores, and insect-eaters and marsupials. Within each section, animals are grouped by family. Articles cover the habits and evolution of each family, and data are provided for the range and distinctive features of each species. The book is profusely illustrated with both excellent color photographs and drawings. A detailed classification chart and an index by common and scientific name are included as appendices. The low price of $45.00 will permit this encyclopedia to be added to virtually any reference collection. For use by audiences ranging from professionals to children (DAT/BL)

Encyclopedia of natural insect and disease control. Edited by Roger B. Yepsen, Jr. Emmaus, PA.: Rodale Press; 1984. 490p. $24.95. ISBN 0-87857-488-3.

Many people are concerned about the larger-scale use of agricultural chemicals; even home gardeners wonder what they can do in their own small gardens. This encyclopedia is a guide for those who wish to control pests without using the standard pesticides. By varying the planting schedule, changing watering techniques or any other of a variety of methods, one can prevent (or at least inhibit) the destruction of crops. It is arranged alphabetically by plant type, and each article discusses the insects, diseases, or even wildlife that affect the plant, and different ways to treat the affliction. There are quite a few line drawings of insects, and over 100 pests and diseases are pictured in color section. Highly recommended for most public and academic libraries and possibly even personal purchase. (KLK)

Encyclopedia of psychology. Edited by Raymond J. Corsini. New York: Wiley; 1984. 4 volumes. $199.95. ISBN 0-471-86594-X (set).

This four-volume set covers areas of psychology that fall within the following subject groups: applied, clinical, cognitive, developmental, educational, measurement, personality, physiological, social, and theoretical. The approximately 2,150 entries are written and signed by experts in the fields and are arranged in one alphabetical sequence. The articles range from 200 to 900 words, with many followed by lists of suggested readings. The subject/name index lists more than 24,000 items so that readers can easily check topics not found under separate entries. Entries for people cover both the living and deceased and give dates of birth and death, place of birth, education, and major contributions to the discipline. A helpful feature is the 15,000 item bibliography listing all citations found throughout the text. The intended audience is wide-ranging and includes psychologists, psychiatrists, social workers, counselors, sociologists, anthropologists, physicians, lawyers, ministers, and the general public. (BL)

(A) Field guide to the moths of eastern North America. By Charles V. Covell, Jr. Boston: Houghton Mifflin; 1984. 496p. $18.95. ISBN 0-395-26056-6. (Paperback: $13.95. ISBN 0-395-36100-1.)

A volume of the familiar Peterson Field Guide Series, this is designed to aid in the identification of moths that have been collected and prepared with their wings spread. Of the 10,500 species of moths known in North America, 59 families and over 1300 species are covered. Inclusion is based on the likelihood of a collector encountering the species in eastern North America. Microlepidoptera are not covered in detail. Each family of moths is described in an introduction to the species within it. Species specific information includes common and scientific names, description, measurements, diet, range, and abundance. Most of the moths that are illustrated are shown life-size. This is a field guide intended for the non-specialist. As such, highly technical terms and microscopic features have been avoided in most instances. Introductory chapters discuss moth anatomy, life cycle, and techniques for collection and preparation. (BL)

Fishes of the world. 2d ed. By Joseph S. Nelson. New York: Wiley; 1984. 523p. $44.95. ISBN 0-471-86475-7.

This expanded version of the first edition (1976) presents a complete revision of the classification of fishes based on new knowledge about relationships among fishes. In addition, more references to recent systematic works are given, more examples of recognized generic names are included for each family, family descriptions are enlarged, and more biological and systematic information is provided. The lower chordates and fishes are treated in a linear order to represent their postulated evolutionary relationships. The author has also included common names and the general range for each family. When known, life history and maximum length of the largest species is given. The reader will find a larger number of useful line drawings in this edition. There are two appendices: a checklist of the extant classes, orders, suborders, and families, and a series of 45 fish distribution maps. The stated purpose of this volume is to provide an introductory systematic treatment of all major fish groups. It will be useful for students and professionals alike. (BL)

Handbook of counseling psychology. Edited by Steven D. Brown and Robert W. Lent. New York: Wiley; 1984. 982p. $60.00. ISBN 0-471-09905-8.

Counseling psychology is defined here "as an applied psychological discipline devoted to scientifically generating, applying, and disseminating knowledge on the remediation and prevention of vocational, educational, and personal adjustment difficulties". This handbook provides a broad survey of counseling psychology with an emphasis on the current state of knowledge and the past and future research trends. The collection of papers, all written by professionals, cover topics such as vocational psychology, assessment, job satisfaction, personal counseling, preventive psychology, the training and supervision of counselors and researchers, and consultations among others. The last section looks at current issues (minority issues, women's issues, etc.) and their implications for the field. Each chapter includes a list of references. (BL)

Library use; a handbook for psychology. By Jeffrey G. Reed and Pam M. Baxter. Washington: American Psychological Association; 1983. 137p. $15.00. ISBN 0-912704-76-4. (Paperback.)

College students in need of an introduction to library research in the field of psychology will find this to be a very useful, easy-to-use source. Concentrating on information available in a typical college library of 100,000 to 300,000 volumes, it offers brief commentary and illustrative minisearches for each type of source discussed. Selected sections treat topics such as defining and limiting a topic, plagiarism, computer searches, citation searching, subject searches, government publications, sources of information for tests and measures, interlibrary loan services, etc. There is a concise explanation distinguishing monographs from serials, monographic serials from periodicals. One chapter deals with the library staff—who does what and how to make the most of their services. All psychology-related major indexes are covered, including *Psychological Abstracts,* ERIC indexes, *Business Index, Biological Abstracts*, and *Social Science Citation Index*, to name a few. This volume will supplement bibliographic instruction programs or will offer solid assistance in lieu of such. (BL)

Longman dictionary of psychology and psychiatry. Edited by Robert M. Goldenson. New York: Longman; 1984. 816p. $39.95. ISBN 0-582-28257-8.

The stated objectives of this reference volume are to provide a comprehensive lexicon with an emphasis on current terms common to the psychosciences and to impart the maximum information in minimal space. The end result is a dictionary of more than 21,000 concise entries weighted toward neurological, physiological, and medical terms. Terms of historical interest and interdisciplinary vocabulary are also treated. It is arranged in one alphabetical sequence. Cross references are used generously. Looking under "mania", for instance, one finds a list of 93 "see also" references to specific manias in addition to a definition. All official terms of DSM-III are included, as is the DSM-III classification scheme (Appendix A). Other appendices list the 488 test entries, the 216 therapies covered, and 177 related fields from which terms have been gleaned for the purposes of the volume. (BL)

Macmillan illustrated animal encyclopedia. Edited by Philip Whitfield. New York: Macmillan; 1984. 600p. $33.65. ISBN 0-02-627680-1.

Described as a "who's who" of the animal world, this encyclopedia covers mammals, birds, reptiles, amphibians, and fishes. With the exception of fishes, it offers a comprehensive listing at the family level of animals within the vertebrate group. Fishes are treated comprehensively at the level of order. Within each family, representative species are discussed with regard to size, range and habitat, lifestyle, breeding habits, and physiology. Scientific and common names are given as well. A key to the conservation status ranges from endangered to indeterminate and is assigned to each species mentioned. An important feature of this volume is the provision of nearly 2,000 color illustrations covering each species discussed. These paintings appear on the page opposite the text, making for convenient consultation. This will be of value to readers ranging from professionals to amateurs. (BL)

Tobacco encyclopedia. Compiled and edited by Ernst Voges. Mainz, Federal Republic of Germany: Tobacco Journal International; 1984. 468p. $50.00. ISBN 3-920615-07-7.

This compact encyclopedia is divided into two parts. The first (and largest) section is an alphabetical arrangement of short definitions, and includes people, companies, foreign language terms, scientific names, and the like. In an unusual twist, color advertisements are interspersed throughout this section, and some of the advertising companies include relevant illustrations, such as types of pipes. The second section consists of short articles arranged in nine broad subject areas. The *Tobacco encyclopedia* is based on the German language *Tabak-Lexikon*, originally published around 1967. Recommended for collections on tobacco production from plant to finished product and even some aspects on tobacco's marketing and consumption. (KLK)

Wildflowers of Arkansas. By Carl G. Hunter. Little Rock, AR: Ozark Society Foundation; 1984. 296p. $29.68 (pap.). ISBN 0-917659-00-7.

The most remarkable feature of this field manual is the photography with its sharp imagery that shows off the naturally vibrant colors of wildflowers. Unter includes descriptions of 484 different wildflowers currently found in Arkansas, each accompanied by a color photograph. However, the collection, though comprehensive, is not exhaustive. Nevertheless, in addition, 116 other flowers are described but without photographs. The author points out that there are, for example, nearly 300 species of sunflowers. This work is intended to complement rather than replace the many field manuals now available. Highly recommended for libraries strong on the natural sciences, especially botany. (KLK)

(The) World's whales; the complete illustrated guide. By Stanley M. Minasian and others. Washington: Smithsonian Bks.; 1984. 224p. $27.50. ISBN 0-89599-014-8. (Distributed by Norton, New York.)

This guide to whales, dolphins, and porpoises covers all 70 recognized cetacean species. It includes nearly 160 color photographs of most of the species swimming in the open seas. It is arranged by suborder, Odontoceti and Mysticeti, and is then broken down by family, genus, and species. The text accompanying the photographs covers physical description, color, fins and flukes, length and weight, throat grooves, baleen plates or teeth, feeding, breathing and diving, mating and breeding, herding, distribution, migration, and natural history notes. Also included is the name and date indicating the person first describing the animal and the year of discovery. An introductory chapter offers basic information about cetacean biology. A glossary and a list of selected readings are found at the back of the volume. (BL)

EARTH SCIENCES

Thesaurus of rock and soil mechanics terms. Compiled by J. P. Jenkins and A. M. Smith, New York: Pergamon; 1984. 62p. $15.00(paper). ISBN 0-08-031632-8.

This thesaurus is a structured list of "controlled" terms (or keywords) used to index the material contained in the *Geomechanics abstracts* database which itself contains bibliographic references to publications pertaining to the fields of rock and soil mechanics. Included are topics such as properties of rocks and soils, mining, tunnelling, foundation engineering, and communication. Terms are arranged alphabetically with preferred ones in bold type and non-preferred terms in italics. For collections on mining and earth sciences in general. Compilers are with the Imperial College of Science and Technology, London. (RGK)

PHYSICAL SCIENCES

Alcohols with water. Edited by A. F. M. Barton. New York; Pergamon; 1984. 438p. $100.00. ISBN 0-08-025276-1. (Solubility Data Series, vol. 15.)

Continues a venerable series on solubility data, with this volume dealing with binary systems containing only water and a monohydroxy alcohol. There are 427 pages of evaluation text and compiled tables. An 18-carbon alcohol is the highest noted. There are three indexes: system, registry number, and author. For all comprehensive chemistry collections. (RGK)

Compendium of thermophysical property measurement methods. Vol. 1: Survey of measurement techniques. Edited by K. D. Maglić and others. New York: Plenum; 1984. 789p. $97.50. ISBN 0-306-41424-4.

This is the first of a two-volume reference set which will serve as a guide in selecting the best technique to be employed in measuring the required property of a substance. This volume deals primarily with thermal and electrical conductivity, thermal diffusivity, specific heat, thermal expansion, and thermal radiative properties of solid materials over a wide range of temperatures. For strong collections on thermophysics. (RGK)

Dictionary of chemistry and chemical technology: English-German. Edited by Helmut Gross. New York: Elsevier; 1984. 714p. $105.75. ISBN 0-444-99618-4.

This volume contains a collection of about 55,000 terms (English to German only) mostly rather specialized for chemistry (some 20 areas) and chemical technology (involving also some 20 subfields in industrial chemistry and chemical engineering). An excellent tool for commercial translators (though possibly a bit on the expensive side for other users). (RGK)

Glossary of technical terms. Surface treatment of aluminum. English-French-German. Compiled by L. Bosdorf and others. Düsseldorf: Aluminum-Verlag; 1984. 358p. $48.00(paper). ISBN 3-87017-163-4. (Distributed by Hayden.)

This is a collection of some 2600 technical terms on the surface treatment of aluminum. Section arrangements in parallel columns are: first, German-English-French; second, English-French-German; and third, French-German-English. For most physical science collections. (RGK)

Handbook of algorithms and data structures. By G. H. Gonnet. Reading, MA: Addison-Wesley; 1984. 286p. $25.95. ISBN 0-201-14218-X(softcover).

This is a work which contains considerable information on algorithms (main emphasis) and their data structures. Thus it is primarily for the programmer who wants to code efficiently but also useful to others who need quick information. One of the appendixes (III) is a compilation of an amazing 683-citation bibliography. Author is with the University of Waterloo, Canada. (RGK)

Handbook of physical and mechanical testing of paper and paperboard. Vol. 2. Edited by Richard E. Mark. New York: Dekker; 1984. 508p. $79.75. ISBN 0-8247-7052-8(v.2).

This volume continues a series on the modern aspects of properties testing in the paper and paperboard field. For those concerned with the planning, specifying, and evaluating of the testing of these materials.

(A) Manual of chemical and biological methods for seawater analysis. By Timothy R. Parsons and others. New York: Pergamon; 1984. 173p. $19.50. ISBN 0-08-030288-2.

This manual will serve to provide, in detail, biological and chemical techniques used primarily by biological oceanographers in their analysis of seawater. The few literature citations provided are imbedded in the text and, unfortunately, there is no subject index. The first author is with the University of British Columbia. (RGK)

McGraw-Hill dictionary of chemistry. Edited by Sybil P. Parker. New York: McGraw-Hill; 1984. 665p. $32.50. ISBN 0-07-045420-5.

This tool focuses on the vocabulary of theoretical and applied chemistry and materials, and includes the specialized terminology of atomic and nuclear physics. The some 9000 terms were selected from the *McGraw-Hill dictionary of scientific and technical terms* (3d ed., 1984). For workers and students in the eleven fields represented. (RGK)

Steam tables in SI-units. 2d rev. ed. Edited by Ulrich Grigull and others. New York: Springer; 1984. 91p. $8.50(paper). ISBN 0-387-13503-0.

These tables and diagrams (though extensive) are mainly meant for use by university students to solve problems in the fields of power and chemical engineering. Nevertheless, the book is also good as a support for industrial practitioners who need a quick and reliable general view of the properties of water substance. Only SI-units are used. (RGK)

SCIENCE, GENERAL

Concise science dictionary. New York: Oxford University Press; 1984. 758p. $22.50. ISBN 0-19-211593-6.

This handy-sized dictionary provides a good general purpose work for high school and college students, as well as non-scientists needing a tool to meet their needs. Emphasis is given physics, chemistry, biology, biochemistry, palaeontology and the earth sciences. In addition key terms are covered as found in astronomy, mathematics and computer science. Scientific jargon is not included. Where needed, a few diagrams or drawings are included. Entries vary in length from a dozen words to half a column, so some entries are quite full. There are adequate cross references. At the price this should provide even the smallest sci-tech collections with a valuable addition to their reference materials. (EM)

Dictionary of scientific and technical terminology. English, German, French, Dutch, and Russian. Boston: Martinus Nijhoff; 1984. 496p. $90.00. ISBN 90-201-1667-3.

Some 9000 entries in English are given in this collection. However, there are also indexes in German, French, Dutch, and Russian which are useful when translating texts from or into any of the languages with which this tool deals. Primarily for engineers working in virtually any branch of science and technology. (RGK)

Russian-English translator's dictionary: A guide to scientific and technical usage. By Mikhail Zimmerman. 2d ed. New York: Wiley; 1984. 544p. $59.95. ISBN 0-471-90218-7.

This expanded edition continues the aim of the original work (1966) by supplying word combinations and expressions for the professional translator so the translation can be free of unidiomatic or amateurish passages. However, it is not a dictionary of terms or idioms but a collection of typical examples from scientific and technical sources. Over 10,000 entries are provided. (RGK)

Science and technology in Japan. By Alun M. Anderson. Essex, U.K.: Longman; 1984. 421p. $85.00. ISBN 0-582-90015-8. (Longman guide to world science and technology, vol. 4.) (Distributed by Gale Research.)

A handy volume which charts the major research programs now (1984) underway in Japan and provides a detailed directory of the research institutes, universities, and industry where they are being carried out. Provides too a comprehensive description of Japan's major government-industry cooperative research projects. Included is an excellent subject index and an alphabetical listing of establishments mentioned in the text. Not only for scientists but also for those active in international trade and investment. (RGK)

TECHNOLOGY

Builder's comprehensive dictionary. By Robert E. Putnam. Reston, VA: Reston Publishing Co.; 1984. 532p. $39.95. ISBN 0-8359-0579-9.

This dictionary is intended to cover all of the technical terms used in the construction industry. Each entry consists of a one-sentence description of the term in question. Definitions correspond to the national standards in the industry and are taken directly from the standards or codes whenever possible. Many of the definitions include illustrations and most have cross-references to related terms. A seperate section of legal, real estate, and management terms is included for those users interested in construction management. This work will be useful for any library with a collection covering the construction industry. (DAT)

(The) Computer and telecommunications handbook. By Jeff Maynard. London: Granada Technical Books; 1984. (Distributed by Sheridan House Inc., Dobbs Ferry, New York.) 237p. $25.00. ISBN 0-246-12253-6.

Much of the information available to those involved in computing or telecommunications is scattered in a wide variety of sources. This handbook provides a single source for an assortment of standards, reference data, tables, symbols, etc. As a reference work it will be an invaluable aid for computer programmers, system designers, business analysts, maintenance engineers, network designers, students, and many others in fields associated with information technology. (CC)

Electroplating engineering handbook. Edited by Lawrence J. Durney. 4th ed. New York: Van Nostrand Reinhold; 1984. 790p. $69.50. ISBN 0-442-22002-2.

This handbook provides a compilation of the most frequently required facts for metal finishers and a guide to additional sources of information. As in the previous editions, the book is divided into two parts: I, General processing data and II, Engineering fundamentals and practice. Chapters have been contributed by experts. While a great deal of the material in previous editions has been retained, some 44 chapters (or subchapters) have either been rewritten completely or have been reviewed and revised. The discussion of design for electroplating has been expanded to include the importance of proper material selection. The chapters on applications and on troubleshooting have been completely rewritten. A chapter on hi-speed plating has been added. (CC)

(The) Encyclopedia of applied geology. Edited by Charles W. Finkl, Jr. New York: Van Nostrand Reinhold; 1984. 644p. $75.00. ISBN 0-442-22537-7. (Encyclopedia of earth sciences, Vol. XIII.)

This volume deals with applied geology which involves the application of various aspects of geology to economic, engineering, water-supply, and environmental problems. 87 main subject entries are used (but with a great number of cross-references). A random review of literature citations for 25 subjects indicated that the most recent date appears to be 1981. On occasion the data is rather outdated as, for example, under "urban tunnels and subways", where the table on page 609 indicates that in Buffalo, NY, a "rapid transit (tunnel is) proposed", a fact taken from a 1976 data source; whereas in reality the tunnel has already been completed and will be in service in early 1985. For industrial and academic reference library collections. (RGK)

Handbook for electronics engineering technicians. 2d. ed. Edited by Milton Kaufman and Arthur H. Seidman. New York: McGraw-Hill; 1984. Mixed pagination. $39.00. ISBN 0-07-033408-0.

This handbook treats fundamental topics in discrete circuits, and also in analog and digital integrated circuits from the point of view of practical applications. As a revised edition, it contains seven new chapters (18-24) covering areas such as microprocessors, active filters, and oscilloscopes. The background for users need be no more than a two-year community college. (RGK)

(The) Handbook of computers and computing. Edited by Arthur H. Seidman and Ivan Flores. New York: Van Nostrand Reinhold; 1984. 874p. $77.50. ISBN 0-442-23121-0.

This handbook is organized into six subject areas: components, devices, hardware systems, languages, software systems, and procedures. It is designed mainly for those workers in the field whose views are becoming rather circumscribed and who need to understand and interact with other computer scientists. Thus the work describes the fast changes in particular areas of computer science. Actually it is an overview of progress in the field. (RGK)

Illustrated dictionary of mechanical engineering. English, German, French, Dutch, Russian. Boston: Martinus Nijhoff; 1984. 422p. $49.50. ISBN 90-201-1668-1.

This very heavily illustrated dictionary is arranged according to 14 sections (branches) of mechanical engineering. The key language is English followed by German, French, Dutch, and Russian equivalents. Over 3600 terms are provided. There are also indexes to all five languages. It is a worthy work especially for students just starting the study of mechanical engineering terms in a foreign language. For engineering collections and even personal purchase. (RGK)

Kirk-Othmer encyclopedia of chemical technology. Index to volumes 1-24 and supplement. 3d ed. New York: McGraw-Hill; 1984. 1274p. $185.00. ISBN 0-471-04154-8.

This general, cumulative index covers all third edition volumes and some 90,000 entries are included. It is a budgetary must for all science and technology collections owning the whole set. (RGK)

Kirk-Othmer encyclopedia of chemical technology. Supplement. 3d ed. New York: McGraw-Hill; 1984. 924p. $150.00. ISBN 0-471-89214-9.

This supplement covers from "alcohol fuels" to "toxicology" with 31 other topics in between (by actual count of the "Contents" but the "Preface" refers to 39 articles). For all science and technology collections which have the 24-volume set. (RGK)

McGraw-Hill dictionary of engineering. Edited by Sybil P. Parker. New York: McGraw-Hill; 1984. 659p. $32.50. ISBN 0-07-045412-4.

This tool represents a collection of 16,000 terms covering some ten areas of engineering but *excluding* chemical, electrical, and food engineerings. Note that the terms were drawn from the *McGraw-Hill dictionary of scientific and technical terms* (3d ed., 1984). For many engineering collections. (RGK)

Metals handbook. 9th ed. Vol. 7: Powder metallurgy. Coordinated by Erhard Klar. Metals Park, OH: American Society for Metals; 1984. 897p. $79.00. ISBN 0-87170-013-1.

This is in reality the 9th edition of the *Metals handbook* but as Vol. 7 of the handbook devoted exclusively to powder metallurgy. Major emphasis is on materials and processes as they are currently used in industry. There are four sections: production of metal powders, characterization and testing of metal powders, and powder systems and applications. For virtually all engineering collections, especially metallurgical, materials, and design. (RGK)

(The) Social and economic impact of new technology 1978-84: A select bibliography. Compiled by Lesley Grayson. New York: IFI/Plenum and Letchworth, Herts, England: Technical Communications; 1984. 80p. $85.00. ISBN 0-306-65209-9 (IFI Plenum); 0-946655-02-2 (Technical Communications).

This wide ranging bibliography brings together American, British, and European literature on the social and economic effects on government, industry, business, and the home. Though certainly not comprehensive (only 700 citations with annotations) it does represent a significant set of references on "new technology" for managers and research workers. An index of authors would have been helpful. The rather high cost of the volume could have been reduced by the use of less elaborate cover boards (in fact flexible covers would have sufficed). (RGK)

(The) Software catalog—science and engineering. Produced from the International Software Database. New York: Elsevier; 1984. 687p. $29.00. ISBN 0-444-00925-6.

This catalog is a buyer's guide to software for sale in the fields of science and engineering. For each program included information is provided on the purpose of the software, the equipment necessary to run the software, and ordering information for obtaining the software. Programs for both mainframes and microcomputers from both commercial and non-profit vendors are included. Entries are arranged by International Standard Program Number (ISPN), which has the advantage of arranging the programs in order by vendor. Indexes are included for the type of hardware required, the operating system used, the programming language used, the microprocessor required, the specific application or subject of the software, and the title of the program. This book should be useful for researchers or managers involved in the purchase of scientific software. (DAT)

Timber designers' manual. By J. A. Baird and E. C. Ozelton. 2d ed. London: Granada Technical Books; 1984. 624p. $75.00. ISBN 0-246-12375-3.

This manual is written primarily for engineers and structural designers, but much of the material included will assist architects and builders in understanding timber components and performance. The authors have limited themselves to discussing the design of the components which constitute the bulk of the work in timber engineering, i.e., mainly beams, columns, and trusses. Both overall and detail design are covered. The book provides many tables of data and coefficients which will save the practicing engineer many design hours. (CC)

U.S. bombers 1928 to 1980s. By Lloyd S. Jones. 4th ed. Fallbrook, CA: Aero; 1984. 280p. $15.95. ISBN 0-8168-9130-3.

Designed as a rather spectacular reference work, it is intended to illustrate the development of the American bomber, step by step, from the Keystone XB-1 biplane of 1928 to the supersonic strategic weapons systems of the 1980s. 75 three-view drawings are used plus frequent insert drawings for clarity, and of course photographs are provided for each aircraft. Some 337 illustrations are included. There is also a chapter on color schemes but the photographs are in black-and-white. For personal purchase and public libraries, plus any collections on the history of aviation. The author is connected with the plastic model industry. (RGK)

What every engineer should know about engineering information resources. Schenk, Margaret T.; Webster, James K. New York: Dekker; 1984. 216p. $24.75. ISBN 0-8247-7244-x.

Serves as a guide to print and nonprint sources of engineering information, aimed at practicing engineers as well as engineering students. Arranged by format, chapters are devoted to such topics as periodicals, technical reports, handbooks, tables, patents, yearbooks, maps and audio-visual materials. Other chapters deal with engineering software, information on licensing of engineers and technical report preparation. The last chapter discusses libraries, information centers and information brokers. Includes use of online databases. The authors cite examples of important works in each category, each of which bears a short annotation, although not listed in the otherwise adequate index. Should prove to be a current, useful source for both users and librarians in technical libraries. (EM)

What's in print—the subject guide to microcomputer magazines. Indexed by W. H. Wallace. Blue Ridge Summit, PA: Tab Books; 1984. 461p. $14.95. ISBN 0-8306-0611-4.

This book is a one-volume index to articles in 75 microcomputer magazines covering 1981-1983. Entries are broadly arranged by machine and are subdivided into 22 subject categories for each machine. A separate chapter is included for articles that are not machine-specific. Within each subject section, the articles are in alphabetical order by title. This index provides only the most basic information for each article, listing only the title, author, magazine, and date (page numbers are omitted). Although it is a very simple index that does not provide a wealth of information about each article, it is still a very good buy. It is recommended for anyone interested in finding information from relatively recent microcomputer periodical issues. (DAT)

SCI-TECH ONLINE

Ellen Nagle, Editor

NATIONAL ONLINE MEETING HELD

More than 4100 registrants attended the 6th National Online Meeting held in New York City, from April 30-May 2, 1985. A record 153 exhibit spaces displayed products and services of 126 exhibitors. This participation represented an increase of 33% in registrants and 25% in exhibitors over last year's record meeting.

The program began with an overview session entitled "Highlights of the Online Database Field—Gateways, Front Ends and Intermediaries." This was followed by more than 80 papers grouped under more than 20 separate topics. Almost 100 product reviews were also scheduled. The program reflected the current state of the online field: 21 papers were presented on end user searching (end users were also referred to as "specialized users" or "novice searchers.") Use of microcomputers and other technology accounted for 19 papers, 4 of which dealt with electronic publishing, and only 2 papers this year were devoted specifically to downloading. Other topics represented on the program included full-text searching and electronic mail. The 521-page *Proceedings* of the Meeting are available for $50 from the meeting organizers: Learned Information, Inc., 143 Old Marlton Pike, Medford, NJ 08055.

DATABASE NEWS

Merck Index Online

The *Merck Index* produced by Merck, Sharpe & Dohme Research Laboratories in Rahway, NJ, is now online through BRS. *Merck Index* is a comprehensive, interdisciplinary encyclopedia of chemicals, drugs, and biological substances. The database includes descriptions of the isolation, preparation, biosynthesis, structure, physical and biological properties, pharmacological actions, toxicity, and medical and non-medical use for more than 10,000 compounds and derivatives.

Chemical compounds are searchable by CAS Registry Number, alternate names, and trademarks. Literature references are provided for most compounds. Other searchable fields include: status of the compound, synonyms, drug code number, trade name, molecular formula, molecular weight, patent country, melting point, lethal dose, physical data, derivative data, use, therapeutic category.

The database is a counterpart to the printed *Merck Index,* 10th edition, 1983. It will be updated every 6 months. The connect hour royalty charge is $20. Offline printing charges and $.25 per citation, online print charges are $.05 for the registry number and literature reference fields; all other fields are free of charge.

PsycINFO Coverage of Conference Papers Enhanced

The PsycINFO database has regularly provided coverage of conference and symposium material appearing in journals and other serial publications. Until recently this material was treated as journal articles with a standard bibliographic citation provided. No information regarding the conference at which the paper had been presented was provided. As an aid to searchers, complete information regarding the conference or symposium will now be included in the bibliographic citation for journals which devote whole issues or special issues to conferences, and for journals which include a collection of conference papers along with regular journal articles.

Thomas Register Online

DIALOG has announced the *Thomas Register Online,* available as File 535. The equivalent to the printed *Thomas Register of American Manufacturers,* this new file provides information on nearly 123,000 U.S. manufacturers, arranged by more than 50,000 classes of products representing over 102,000 trade or brand names. Produced by Thomas Publishing Company, the file includes the current 1985 listings for more than one million individual product and service sources.

Records include company information (name, mailing address,

and telephone number) as well as product information (searchable by name or subject term). Selected records also display information on officer names and titles, asset rating, number of employees, parent companies, subsidiaries and divisions, and an indication that a company is an exporter.

The database has approximately 127,000 records and will be reloaded annually. The price for searching the file is $100 per connect hour. Print costs are $1.50 per full record printed online or offline.

NLM Announces CHEMLINE Regeneration

The National Library of Medicine's *CHEMLINE* (chemical dictionary) file was regenerated in April 1985 with several important new additions. The number of drug names in the file has been increased and the quality of nomenclature has been strengthened. Ring data fields have also been updated. The new file includes names taken from the *CAS Index Guide,* as well as from NLM's *RTECS, TDB,* and *MESH Vocabulary* files. In addition to these sources of names, additional drug names were obtained from United States Pharmacopeial Convention Inc. (USP). Over 60,000 records have been augmented with more than 120,000 names and synonyms to increase the user's chance of finding substances of interest. In order to remain current with the names of substances that appear in the popular press but do not appear in *CHEMLINE's* normal sources, staff at NLM have monitored the newspapers and trade literature for chemical names of high interest to be included in the database. These names have been added to *CHEMLINE* if they do not otherwise appear in the file. *CHEMLINE* serves as a central point for chemical identification and as a locator for chemical information in other NLM files.

OCLC Subset Available

BRS has announced an agreement with OCLC Online Computer Library Center to provide a subset of the OCLC Online Union Catalog online from BRS for subject searching through the OCLC EASI Reference Service. OCLC EASI (*E*lectronic *A*ccess to *S*ubject *I*nformation) Reference will enable BRS subscribers to conduct full-text searches for approximately one million bibliographic records (without holdings information) from the OCLC Online Union

Catalog. The OCLC EASI Reference database is an extensive online file of books, serials, sound records, scores, audiovisual materials, manuscripts and software. The materials included in the file have imprint dates within the last four years. Access to the records will be through numeric and title searches as well as keyword and controlled vocabulary (Library of Congress Subject Headings and Medical Subject Headings) searches. The full-text search capabilities of BRS including full Boolean searching are available; all BRS subscribers including both OCLC member and non-member libraries may search this database.

Pharmaceutical News Index Expands Coverage

Coverage in *Pharmaceutical News Index* produced by Data Courier has been expanded with the addition of 5 publications. The database which contains information on international veterinary publications, genetic engineering, biomedicine, and the cosmetics industry is available from DIALOG as File 42.

The new publications which have been added are the following: *Animal Pharm World Animal Health News* which offers international coverage of company and product news in veterinary pharmaceuticals. Agriculture and livestock information along with statistical data for various countries are reported. Market analyses, regulations and laws, R&D studies, and product introductions and approvals are featured. *Applied Genetic News* features news about genetic engineering developments and companies. Major areas include medical, pharmaceutical, and agricultural applications of genetic engineering technology. *Biomedical Business International* covers biomedical news and trends worldwide. It includes current information on medical devices, instrumentation, and imaging. Market analyses, acquisitions and divestitures, and news about privately-held companies are also featured. *FDC Reports—"The Rose Sheet"* contains up-to-date information on the cosmetics industry. News is focused on companies, people, and products in toiletries, fragrances, and skin care. Laws affecting the industry, policies, and regulations of the U.S. Food and Drug Administration, as well as other government agencies are reported. *The Rose Sheet* also features the Weekly Trademark Review, an overview of all marks registered and published for opposition, as compiled from the *Official Gazette* of the U.S. Patent and Trademark Office. *Health Devices* is published by the Emergency Care Research Institute for its membership. It presents comparative evaluation of medical devices by brand and model. *Health Devices* includes reports of deficiencies and hazards based on extensive investigations by equipment manufacturers and users.

PUBLICATIONS AND SEARCH AIDS

PsycINFO Thesaurus

The 1985 *Thesaurus of Psychological Index Terms* is now available. This 4th edition of the *Thesaurus* has been updated, revised and expanded. New features include approximately 250 new terms, 400 new scope notes (in addition to the 1300 previously included), and an enhanced Relationship Section. The *Thesaurus* includes more than 4500 index terms used to index documents in the *PsycINFO database* and the printed *Psychological Abstracts.* The Relationship Section is a hierarchical listing of all terms. The section now contains the number of postings in *PsycINFO* for each term (through June 1984); beginning dates of usage for all terms within a hierarchy; and subject term codes for online searching. Other changes in the publication include revised hierarchies, 160 additional cross references, and an improved organization of sections. As in previous editions, the Rotated Alphabetical Term Section provides permuted access to subject headings. The 1985 *Thesaurus* may be ordered from: American Psychological Association, Order Department, 1200 Seventeenth Street, N.W., Washington, DC 20036. The cost is $45 ($36 for APA members), plus $1.50 shipping and handling charge.

Zoological Record Search Guide

BioSciences Information Service (BIOSIS), producer of *Zoological Record* has announced publication of *The Zoological Record Search Guide* to aid in searching the oldest and most comprehensive index to the world's zoological literature. The *Search Guide* features a Master Index consisting of all controlled terms used in the printed *Zoological Record,* volumes 115-119, covering 1978-1982. The Master Index which includes cross references to preferred terms has more than 15,000 entries. Separate Subject and Systematic Indexes provide hierarchical listings. Content and Profile Guide

Sections provide details about the scope of the database, taxonomic nomenclature rules, and techniques for developing effective search strategies. Priced at $50 per copy, the Search Guide is available from BIOSIS User Services, 2100 Arch Street, Philadelphia, PA 19103-1399. Telephone (800)523-4806.

P/E NEWS Index Guide/Keyword List

This new search aid for *P/E News* is available for purchase from the American Petroleum Institute. The *Guide* consists of: a summary of the rules and guidelines reflected of the API indexing philosophy and procedures; an Index Guide which is an alphabetical listing of keywords with scope notes and cross references; and a Keyword List which provides a complete alphabetical listing of all keywords used and their frequency of use. The *P/E News Index Guide/Keyword List* is available for $35 per copy. Contact: American Petroleum Institute, Central Abstracting & Indexing Service, 156 William Street, New York, NY 10038.

Cambridge Scientific Abstracts Manual Revised

Cambridge Scientific Abstracts has published a newly-revised manual, the *Cambridge Scientific Abstracts Online Users Manual.* The manual consists of 6 main sections describing the 6 databases which comprise the CSA online files: *Aquatic Sciences and Fisheries Abstracts, Conference Papers Index, ISMEC, Life Sciences Collection, Oceanic Abstracts,* and *Pollution Abstracts.* The manual includes detailed sections covering sample records, search strategies, complete subject code listings, formats and costs. The complete manual for all 6 databases is available for $50. Individual database sections may be purchased for $20 each. To order, contact: Cambridge Scientific Abstracts, 5161 River Road, Bethesda, MD 20816. Telephone: (800)638-8076.

SCI-TECH IN REVIEW

Suzanne Fedunok, Editor

DO ENGINEERS READ?

Nkereuwem, Edet Efiong. *An analysis of information use by scientists and engineers in the petroleum industry in Nigeria.* Ann Arbor, MI: University of Michigan, 1984. 195 pp. Dissertation. DA 8422299.

The author found, as a result of a survey questionnaire of 324 scientists and engineers, that there was no significant difference between the scientists and engineers in the frequency of use of published information sources, that the engineers used interpersonal sources more. There was no significant difference in the effectiveness of published and interpersonal sources in meeting the information needs of both groups. Scientists used libraries more frequently than engineers. Finally, it seemed there was no absolute correlation between frequency of library use and perception of users of the adequacy of libraries in meeting their information needs.

SCIENCE FOR THE NON-SCIENTIST

Fjallbrant, Nancy; Sjostrand, Brita. Bridging the gap: library user education to new user categories. *Tidskrift for Dockumentation.* 39(4): 106-110; 1983.

The authors, who are librarians at the Chalmers University of Technology in Gothenburg, Sweden, report on the efforts of that library to cope over a twelve-year period with use of the library by a growing number of the general public seeking information on scientific and technical topics. Brief descriptions are given of programs for secondary school children on the use of chemical literature and computer programming, the use of engineering literature for adult education students, introduction to research techniques for nurses and physical therapists, offshore technology, marketing information for industrial users, and online searching for journalists and television producers. (4 refs.)

COMPUTERIZED REPRODUCTION OF ENGINEERING DRAWINGS

Sondergrath, Michael. Improved engineering expectations with active microfilm aperture card system. *Journal of Information and Image Management.* 17(9): 36-38; 1984 September.

Mr. Sondergrath is manufacturing specifications manager in the Logic Systems Division of Hewlett-Packard Company. In this paper he describes how a collection of thousands of engineering drawings in vellum master copies were converted from a passive to an active microfilm system using aperture cards. The "communication by card" system is described in some detail. The system resulted in improved drawing preservation and protection, in better quality print copies, and in more efficient reproduction. "Since installation of the new system, monthly production of prints has nearly doubled from 15,000 to over 27,000 copies." (no refs.)

CURRENT AWARENESS BULLETINS

Green, Kevin E.; Whiting, Joyce. Combined production of a current awareness bulletin and database on a microcomputer. *Programming: Automated Library and Information Systems.* 18(4): 298-307; 1984 October.

The authors are the Head and Deputy Head respectively of Information and Library Services for the GEC Engineering Research Center in Leicester, England. In this paper they report on the library's use of a Torch C-68000 microcomputer to create an inhouse database of articles from current journals. The database is then used to produce two weekly current awareness bulletins for the company, using a word processor and Wordstar, Superfile and MailMerger software. The programs used are given in a table in the text of the paper. (3 refs.)

COMPUTERIZED AEROSPACE LIBRARY

Montgomery, Alan. CAIRS in use: operation at Bristol Aerospace Dynamics Group, Bristol. *Program: Automated Library and Information Systems.* 18(3): 221-230; 1984 July.

Mr. Montgomery is the Head of Library Services at British Aerospace in Bristol, England. After giving some background on the library, which has a staff of twelve and a collection numbering over 80,000 technical reports, 400 periodical subscriptions, and approximately 5000 monographs, the author explains why the library decided to install a computer-assisted information retrieval system. The list of specifications drafted in 1977 by the library staff is included in the paper, which continues with information about the early development of the system, implementation, and concludes with a description of the present configuration of equipment and software programs. The paper ends with the following advice in the form of "Lessons":

> Very careful planning of the database library and file structures was time well spent with a system like CAIRS. . . . A good system suggests its own development and if most of the important extras are already available, further growth of the system is not slowed down by waiting for additional hardware or software. It is easy to underestimate the number of terminals needed and the amount of processing power required. . . . Maintenance and servicing contracts are essential . . . Finally, staff must be thoroughly trained and enthusiastic about the system. (3 refs.)

INDUSTRIAL LIBRARY COOPERATION

McElroy, A. Rennie. Academic/industrial links. *Information and Library Manager.* 4(2):40-43; 1984.

The author is Deputy Librarian at Napier College in England. In this paper, which was first delivered as a speech, he concentrates on the obvious advantages to be gained by cooperation between company libraries and those in academe, stressing economy, service to users, and staff development. He concludes with the comment that "the extent to which interaction can occur between colleges and the employment sector is influenced by senior management attitudes and philosophy, but the most significant factor by far is the dynamism—or inertia—of individuals. . . . Specialists are appointed to advise, inform, educate, and to take or to influence decisions: librarians should accept that opportunity; only then will the scope and variety of interactions between college and company approach its potential." (no refs.)

Roaf, Margery. The industrial library and the information network. *Information and Library Manager.* 4(2): 247-251; 1984 December.

In the same issue of *Library and Information Manager,* the Information Officer of Foster, Wheeler Power Production Company in the U.K. continues the topic of library cooperation by describing such formal links as interlibrary loan and the improvements lately implemented by the BLLD. After mentioning other links such as union catalogs, periodical lists, and the information to be gained from visits to other libraries, she goes on to describe the more important element of "free floating goodwill" which enables special subject librarians to override the complexity of access to information in government and business. She concludes, "I see much of the success of the industrial units lying in their ability to tap outside resources, and their knowledge of other library networks. . . . This must not be lost in the welter of high technology and half formed commercial attitudes." (no refs.)

EPONYMS

Diodato, Virgil. Eponyms and citations in the literature of psychology and mathematics. *Library and Information Science Research.* 6(4): 383-405; 1984 October-December.

In this paper the author, an Assistant Professor at the School of Library and Information Science at the University of Wisconsin-Milwaukee, proposes that eponyms (an expression that consists of an individual's name plus a word denoting some idea or thing associated with that person) may be an important tool in database searching in the fields of psychology and mathematics, where eponyms are not uncommon. Surveying 4506 articles published in 1982, he discovered that 4.4 percent of psychology titles and 33.4 percent of mathematics article titles contained at least one eponym each. The hypothesis that eponyms may point to related literature seems to be borne out by the result that in psychology 74 of 95 eponyms occurred in articles that also cited the same names in their bibliographies. In mathematics the result was 685 of 1105 eponyms names were cited in the bibliographies. The eponyms also matched index entries for those articles in 16.8 percent of the psychology articles and in a much larger 39.9 percent of the mathematics papers. The author concludes with some restrictions on the use of this technique for literature searching. (9 tables; 22 refs.; 6 appendices)

Aversa, Elizabeth Smith. *Citation patterns of 400 scientific papers and their relationship to literature aging.* Philadelphia, PA: Drexel University, 1984. 191 pp. Dissertation. DA 8412612.

The author looked for patterns in the papers studied, and for relationships among any patterns and the aging rates of the papers studied. First, the citation histories of four hundred papers published in 1972 were plotted. One pattern, characterizing 42 percent of the papers, was a high early increase in the number of citations, followed by a rapid decrease in citedness after the third year. Fifty-eight percent of the papers fit into a second pattern, of a slower takeoff period, reaching a peak in the sixth year after publication, and maintaining a higher number of citations subsequently. The author believes that there are also two non-age-related-variable relationships between the two above citation patterns and either the number of authors of the paper, or the number of times the author of a paper was cited for earlier papers.

WHAT PRICE SYMPOSIA?

Watkinson, Anthony. Meeting in print. *Nature.* 312(5991): 201-202; 1985 January 17.

In this short essay introducing the "Autumn Books" supplement of *Nature* magazine, Mr. Watkinson, who is a Commissioning editor at Oxford University Press, reports that science publishers do not heavily promote the symposia on their lists, but that such publications often account for over one quarter of the list of any major scientific-medical-technical publishing house. Symposia do make money for publishers, largely because they are produced from camera ready copy, the editorial work is not done in-house, and because royalties are lower than the standard fifteen percent, if they are paid at all. Finally, Mr. Watkinson admits that a major factor in the profitability of symposia is that "it is generally agreed that symposia are not price-sensitive items. In theory publishers price to a market. . . . In practice, although types of books mainly bought by libraries may be priced in accordance with arcane formulae that do not relate to costs—bringing into play such concepts as price ceilings—publishing editors fight the rules to keep down the price of their own creation. . . . No one cares about symposia, especially the camera-ready variety. From some houses amazingly high prices and absurdly low breakevens are par for the course." He concludes that this is the publisher's point of view, and that from the scholar's side, there are doubtless valid reasons for some small number of symposia to see the light of day. (no refs.)

The Use of TeX in MathSci: A Way of Solving the Problem of Display of Special Symbols

Patrick D. F. Ion

The presence of special symbols in databases has long been a problem for database producers and vendors. End-users would like to see the sorts of formulary to which they are accustomed in their work, but the ordinary terminal cannot display it.

MathSci, the database of mathematical sciences produced by the American Mathematical Society, is a case in point. The bulk of the records there are derived from the online version of the journal *Mathematical Reviews.* The journal contains reviews and abstracts of the whole world's mathematical literature and thus includes mathematical formulas and symbols of all types. With the start of 1985, *Mathematical Reviews* changed to the use of a new typesetting system called TeX (TeX is a trademark of the American Mathematical Society) which helps solve the problems mentioned.

TeX was developed with particular attention to the needs of mathematics by the Stanford mathematician and computer scientist D. E. Knuth. Not only does the system set mathematics well with relatively little trouble, but it also is remarkably flexible and has been used for many other purposes. In particular TeX is well-suited to the typesetting of material extracted from databases. The different fields of the records can be tagged and treated in different ways typographically. The special symbols are encoded in the form of an escape character, which is conventionally a backslash (\),

Patrick D. F. Ion is Associate Editor, Mathematical Reviews, 416 Fourth Street, P.O. Box 8604, Ann Arbor, MI 48107.

followed by a string of any number of letters. The string is considered to be finished when anything other than a letter, such as a space, say, is reached. Thus int is the code for an integral sign and bullet that for a printer's bullet (big black dot). TeX encoding is readable in simple cases and decipherable in all cases, and it is also fully transportable since it contains no non-printing control characters. Since *Mathematical Reviews* is typeset using TeX, it is natural to provide as the online version of the reviews the TeX code for the typeset material. This can be displayed online and contains the full information, formulas and all.

Just recently, in March 1985, an exciting new development occurred. The whole typesetting system TeX, which was heretofore only available as public domain software on mainframe computers, can now be bought in an implementation for the IBM-PC/XT or AT. Now it is possible to do a search of the newest material in MathSci on DIALOG, say, to retrieve the text and capture it with appropriate communications software, and then to serve this up to TeX. This will produce a file which can be printed on an appropriate graphics printer attached to the PC.

The result is typeset mathematics at a low but quite acceptable resolution, as the end product of a DIALOG search. The end-user can again see the formulas. Indeed, on an Apollo or Sun workstation one can even display the document on a high-resolution screen as though printed there, and there is talk of such a facility for the AT.